U.S. Agricultural Research:

Strategic Challenges and Options

edited by
Robert D. Weaver

Agricultural Research Institute
Bethesda, Maryland
March 1993

Published in March 1993 by
Agricultural Research Institute
9650 Rockville Pike
Bethesda, Maryland 20814
(301)530-7122

Stan Cath, Executive Director

The views expressed in this publication are those of the authors. This publication is part of the overall program of the Institute, as endorsed by its Board of Directors, but does not necessarily reflect the views of individual members of the Board or the Advisory Committee.

Printed in the United States of America

Library of Congress Cataloging in Publication Data

U.S. Agricultural Research: Strategic Challenges and Options

1. Agriculture - Research. 2. Research Policy

I. Weaver, Robert D.

ISBN #: 0-944919-04-9

Printed and bound in the United States of America.

Preface

During the past decade, the evolution of the U.S. agricultural research system has been affected by significant internal and external forces. Recognition of these forces has resulted in both reactions and attempts to plan for adaptation to anticipated changes. Recent Federal budget cycles have seen substantial efforts by both public and joint public and private sector groups to influence and determine the ag research agenda, the levels of funding, as well as means of allocation of funding for at least public sector agricultural research. These past efforts have focused on strategic planning and agenda specification to identify near-term strategies appropriate for near-term environments. While these strategic planning efforts were influenced by private sector and interest group comment, they did not have as their goal an assessment of the system's adaptation and performance as forces have changed over the longer-term.

The goal of this book is to contribute to such a strategic assessment of the U.S. agricultural research system. With this in mind, its chapters focus on the following objectives:

1) Enumeration and description of key forces which are affecting the allocation of scientific effort in the ag research system,

2) An examination of key institutional responses to these new forces,

3) An assessment of the implications of these institutional responses for the performance of the agricultural research system, and

4) Development of an agenda of strategic issues and possible actions which could result in improved performance of the U.S. ag research system.

The papers included in this book are drawn in part from presentations made at a two day international symposium on the Dynamics and Performance of the U.S. Agricultural Research System held in McClean, Virginia September 13 - 19, 1992. The symposium was organized by the Agricultural Research Institute in collaboration with sponsorship by and support of the Africa Bureau of the U.S. Agency for International Development, the Special Program for African Agricultural Research, the U.S. Department of Agriculture, the World Bank, and the Pennsylvania State University. The symposium program was developed through an organizing committee. Throughout this process, the support of Stan Cath, Executive Director of the Agricultural Research Institute, and his staff were invaluable.

Robert D. Weaver
Pennsylvania State University

Organizing Committee
Dynamics and Performance of the U.S. Agricultural Research System

Robert D. Weaver, *Chairperson*, The Pennsylvania State University
C. Eugene Allen, University of Minnesota
Lawrence Busch, Michigan State University
Will D. Carpenter, Monsanto Agricultural Co.
Donald N. Duvick, Pioneer Hi-Bred International, Inc.
Robert E. Evenson, Yale University
Robert M. Goodman, Calgene, Inc.
Theodore L. Hullar, University of California, Davis
Paul O'Connell, CSRS/LISA Program
Vernon Ruttan, University of Minnesota
James J. Zuiches, Washington State University

Preparation of the manuscripts was guided by the expert hand of Ms. Barbara Gervinski.

Table of Contents

Introduction

Chapter 1

The U.S. Agricultural Research System Under Challenge

Robert D. Weaver
Pennsylvania State University

The 1980's brought substantial changes in the focus and scope of agricultural research demanded by society. At the same time, the scientific opportunity with potential relevance for agricultural applications expanded dramatically. Together these fundamental changes in demand and supply resulted in challenges to the U.S. agricultural research system to adapt. The expansion of the scope of demands placed on the agricultural research system was enormous. In one dimension, these demands expanded from those for new farm production technology to encompass the full realm of agricultural production and food processing all the way to the consumer's dinner table. In another dimension, demands for agricultural research expanded from those for enhancement of farm productivity and profitability to improving the *external* impacts of agricultural and food system production and marketing activities. External effects, or externalities, are outcomes or effects of an economic activity that occur outside the scope of interest of the manager of the activity. Broad social interest in externalities has followed naturally from increasing recognition of the costs and benefits generated by these effects of production, processing, and marketing activities. Of greatest concern have been external effects such as water and air quality, food quality and safety, nutritional quality and variety, and waste from food production activities.

The expansion of the scope of demands on the agricultural research system (vertically through the food production and marketing system, as well as for the consideration of externalities) has far exceeded the expansion of the supply of agricultural research. The result can be viewed as a substantial excess demand for the services of the traditionally farm based agricultural research system. Both the scale and scope of this excess demand has challenged the U.S. agricultural research system to look beyond tradition and the status quo to find new means for response.

Changes in Political Support for Financing of Public Sector Research

As the scope of the demands for and potential supply of agricultural research system have expanded, political and public financial support of agricultural research has weakened for three reasons. First, the farm client base continued to shrink, a process induced by technological change and associated expansions of productivity and the scale of farming. This reduction in the farm client base relative to expanding urban constituencies directly reduced the voting power of interest groups which could identify direct and appropriable benefits of agricultural research. This traditionally solid political base for agricultural research was replaced by a highly diversified base for which the benefits of agricultural research are relatively indirect and often intangible. This new, highly diversified constituency for agricultural research includes both domestic and foreign producers, marketing system participants, consumers, and interest groups concerned with specific externalities associated with the agricultural and food industries. A second reason for reduced financial support of agricultural research has followed from reduced public budgets which have placed research funding in heated competition with other demands for public resources. Third, the private sector's role in application and diffusion of the results of agricultural research expanded and matured, weakening the direct linkage between public agricultural research and farm clientele. The resulting increase in political distance between public research sources and farm clientele weakened the political support for public agricultural research within the farm community.

The shrinking of traditional bases of political support, combined with diversification into new, diffuse, and weaker bases of support, has resulted in less effective political support for public agricultural research. A further force for diversification of scope has resulted from an increasing recognition that the public agricultural research and extension systems could effectively generalize their sectoral focus to include non-agricultural issues and clientele groups which are perceived as being serviceable by the current system's resource base. To the extent that these activities complement agricultural research and are fully financed, an increase in the productivity of the agricultural research system has been expected.

Induced Change in the U.S. Agricultural Research System

The changes in the demands for and potential supplies of agricultural research have induced substantial changes affecting the interaction between demand and supply of research. Changes in the scope of demand have resulted in at least three substantial changes in the way in which demands are expressed or communicated to the agricultural research system. First, new constituencies have called for regulatory coercion of the research system to ensure their demands are met. Typically, though not exclusively, these demands of regulations have focused on externalities of research or commercial applications of research products. In all cases the changes have altered the incentives for both the practice of and products of agricultural research. Second, constituent groups have increasingly demanded the centralization, rather than decentralization, of the processes of research agenda formation. This change is evident both within the context of public budget formation and in the evolution of political action agenda for regulation targeted at both private and public research. Importantly, the shift to centralization of the research agenda reflects a third change: a fundamental shift in the franchise for agenda setting from the scientist to social interests. Implicitly, these changes imply the emergence of a call for a broad-based reassessment of the adequacy of the institutions and processes through which the agricultural research system is managed.

In parallel to these changes induced on the demand side, substantial changes have been induced by the emergence of new opportunities in basic sciences on the supply side of the agricultural research system. First, substantial changes in incentives for both private and public sector research have resulted from changes in intellectual property rights induced, in part, by new opportunities for agricultural research. Affected changes have included direct changes in intellectual property rights (IPRs) as defined by patent, copyright, and right-to-know or other information disclosure laws. Despite these changes, substantial concerns remain over IPR issues which remain unresolved such as international harmonization, dependency among genetic innovations in plants or animals, and adequacy of patent term given the time requirements of regulatory processes.

An alternative to IPRs as a means of preserving incentives for research is the reservation of the technical nature of innovations as trade secrets. Where IPRs have been judged inadequate, the changes in both demands for and the potentials for supply of research solutions have resulted in changes in institutional forms and roles in the research process, in part targeted at maintaining protection of innovations through trade secrets. One such change in institutional roles has been a shift of effort and expenditure from the public to the private sector. While the determinants of this shift are not well understood, changes in incentives and scientific opportunity have likely played critical roles.

Changes in the Performance of the System

The implications of these changes affecting the U.S. agricultural research system continue to evolve. However, several implications may be cited as potentially resulting in changes in the U.S. agricultural research system's performance. First, both public and private sector directions or lines of research have diversified into new agricultural and off-farm effects or externalities issues. Agricultural research, once narrowly focused on specific opportunities for enhancing farm productivity and profitability, now takes a broader focus encompassing entire production systems often requiring multidisciplinary collaborations. At the same time, public agricultural research has begun to apply resources to issues which go beyond the farm gate. New scientific opportunity has shifted both the private sector's roles in basic and applied research and drawn nonagricultural scientists to consider issues with agricultural implications or application.

These changes in forces impacting U.S. agricultural research activities have also impacted the financing of the system. First, diversification of the focus of research has increased competition among research issues for the available resource base. Second, in both private and public research, these forces appear to have led to an increasing need to demonstrate applied pay-offs to constituencies/markets to ensure that funding has paid-off. A corollary to this implication is that financing of long-term, basic research necessary to fuel future opportunities for application is more difficult to justify to funding sources. Finally, while the forces affecting the agricultural research system have resulted in a diversification of focus, the scale of finance in many areas may not have kept up, leaving many areas of the U.S. agricultural research inadequately financed both absolutely and relative to other national systems.

Changes in the performance of the agricultural research system can be expected as a direct result of the forces affecting the system and, as an indirect result, of changes in institutional roles, focus and scope of, and modes of participation in the system. Several bases exist for assessment of the system's performance. Nonetheless, a traditional measure of performance remains paramount for both private and public efforts - the system's ability to select and resolve issues which maximize the net benefit to society. Achievement of maximum performance according to such a criterion requires a system that is at once flexible and responsive to evolving issues, as well as sufficiently committed to lines of research to provide a stable base of resources and knowledge from which solutions can emerge.

The Need for Strategic Assessment

The current challenges faced by the U.S. agricultural research system have jointly placed the system under close scrutiny by constituents placing demands on the system or competing for its resources as well as by scientists and researchers, both within and outside of the system. Within this context, the imperative for a broad-based, objective strategic assessment of the performance of the system would seem unquestionable. At the same time, such a strategic assessment goes well beyond the scope of the resources available for this book. Instead, this book attempts only to contribute to such a process.

The remainder of the book is organized into three sections. In the next section, *The Current State of U.S. Agricultural Research: Three Views*, interpretative views of a former State Agricultural Experiment Station (SAES) administrator and two long time students of the agricultural research system are given. These chapters are followed by a section composed of chapters which present an empirical view of the research system. Evidence is presented to assess the current state of the U.S. system, in general, and the roles of the private and public sectors, of the SAES system, and of the markets for scientists and human resources, in particular. Together, these two sections provide both an interpretative and empirical assessment of the U.S. agricultural research system not available elsewhere.

The next two sections of the book contain chapters which, respectively, assess two specific public sector responses to changes in demands for and supply of agricultural research: 1) regulation and 2) intellectual property rights. In each case, these responses have emerged as a result of political actions of constituents of the research system on both the demand and the supply side. Importantly, the two responses considered represent different approaches to satisfying specific aspects of demand for agricultural research. The final two sections of the book assess another institutional response to changes in the market for agricultural research: changes in financing and funding mechanisms and strategies by the private and public sectors. A concluding chapter summarizes the preceding chapters and presents an attempt at the identification of key strategic issues and options raised.

The Current State of U.S. Agricultural Research: Three Views

Chapter 2

U.S. Agriculture Research: An Assessment of the System

Walter R. Woods
Kansas State University

Agricultural research has made significant contributions to U.S. and world agriculture by improving efficiency of production, assurance of a high-quality and safe food supply, addressing environmental issues, and conservation of natural resources. By seeking solutions to problems as well as identifying new technologies, publicly supported research has played a large part in the development of U.S. agriculture. Thus, one has to be very positive about the importance of agricultural research based on its past accomplishments, but also based on its potential to contribute to solutions to future problems.

The question of where U.S. agricultural research is going is complex. It is complex, in part, because of the many participants involved in developing and supporting U.S. agricultural research policy. Participants include federal and state agencies, land-grant universities, scientific societies, private companies, cooperatives, foundations, and commodity and general farm organizations as well as Congress and legislatures who appropriate federal and state funds. A second component of this question deals with the uncertainty of commitment to the importance of agricultural research by decision makers at the state and federal levels as well as by commodity and general farm organizations. A third component is the large number of organizations and individuals, not typically associated with agriculture, who are involved in shaping agricultural policy, regulations, and research objectives. Many are attempting to shape policy by the single-issue approach. The fourth component of this equation is the unprecedented rate of change occurring in the U.S. and the world in relation to agriculture. This fourth component has become a major factor because many of us have not adopted policies and processes that allow us, individually or as organizations, to respond quickly enough. Research cannot be turned off or on at will and be successful. It must be viewed as a commitment for the long term in order to have the desired payoff. It is this long-term commitment that has helped keep U.S. agriculture competitive and strong. The expectation is for agriculture to be profitable. It is emphasized that contributions from research are only one of a number of factors that influence the profitability of American agriculture.

Current Status of Agricultural Research

A quick analysis would indicate that U.S. agricultural research is in solid shape because of the strength of U.S. agriculture. However, surface analyses are sometimes misleading; thus, this paper will analyze the question from other points of view. U.S. agriculture operates in a global environment, which increases the importance of U.S. research in support of American agriculture. The importance of agricultural research is questioned by some decision makers today because of continued food surpluses, however, if U.S. agriculture is going to remain competitive, new technology that enhances the potential for profitability of U.S. agricultural industries must be generated and applied. This paper will not analyze how specific sectors of agriculture may fare with respect to increased or decreased research. It is believed that all segments of U.S. agriculture would benefit in their international competitiveness from increased agricultural research. Further, it is understood that research is only one of the factors that will affect the profitability and competitiveness of U.S. agriculture in the international marketplace.

There are several issues that shape the environment that influences support of U.S. agricultural research. These include the past successes of agricultural research, maturity of the industry, global competition, and fiscal capacity, to name a few. U.S. agricultural research has been very productive and effective in helping U.S. agriculture. This effectiveness has lulled some people and organizations into feeling that the research pipeline will continue to flow with information and new technology regardless of what happens to the agricultural research infrastructure and capability. Replacing obsolete equipment, modernizing laboratories, obtaining new technology, and supplying technical assistance to support cutting-edge research are not occurring at the rate needed. In fact, scientist effort at many land-grant universities is on the decrease.

U.S. agriculture is a mature industry, as defined by the fact that the supply exceeds the demand and the opportunity for growth in the U.S. is limited. Certain niche markets, value-added or commercialized agriculture products, may offer growth opportunities. On the demand side, research directed toward meeting new consumer demands could expand demand. The federal government and many state governments are emphasizing value-added products or commercialized products from agricultural commodities to create increased demand for U.S. agriculture products. Value-added research is important to U.S. agriculture not only for local consumption or use, but for export as well. This commercialization has created the feeling in some circles that it is the solution to our problem of excess agricultural supplies; however, it must be emphasized that this is only one of a number of alternatives that need to be considered. As new markets are created, there is still a critical need to emphasize research in support of efficiency of agricultural production, quality of products produced, nutritional value, etc. From another perspective, U.S. agriculture is in global competition for many of the markets that historically have been major consumers of U.S. products. Given the U.S. government's role in exports, its policies may be critical factors in determining the international demand for certain kinds of commodities and products.

A further element of the environment affecting agricultural research is the resistance to increased taxes that has resulted in many states in reduced support of their land-grant universities, thus reducing support for public sector agricultural research. The growing number of states that have reduced support for their land-grant universities adds a vital new dimension in our assessment of the current environment for agricultural research. Federal support through base funds has not kept pace with state funding for a number of years and the federal-state partnership for supporting public agricultural research has eroded.

From a political perspective, agriculture represents a minority of the population and is fragmented into special interest groups. This fragmentation has left many feeling that farm and commodity organizations are only interested in supporting research if it specifically assists their special interests. However, the transferability of information makes it important to provide encouragement for a base level of research support. A further element of the political environment has been the anti-science and anti-chemical movement which has argued that new technology may be bad rather than good. This negative view of new technology places agriculture at a disadvantage if other countries are supporting technology development at a higher relative rate than does the U.S.

A final element of the political landscape is the emergence of environmental issues such as water quality and quantity reflect concerns of citizens and special interest organizations. These concerns have placed agriculture in the role of villain because of its widespread use of agricultural chemicals. The backlash from this perspective has come to view agricultural research negatively, rather than positively. Current agricultural research to reduce pesticide use, cultural practices to control runoff, drought resistant varieties, as well as many other research areas are contributing to improved environmental quality and conservation of our water resources.

Other issues could be mentioned under the description of "our agricultural environment today," but the issues reviewed above are sufficient to emphasize the point that significant pressure is on the public agricultural research establishment. Land-grant universities and USDA are the primary public entities that are doing public sector agricultural research. The role of private companies in agricultural research is not addressed in this paper, yet it is important to emphasize that the private research establishment plays a key role in support of U.S. agriculture. Applications from private research may or may not target a special area of need within the U.S., but will be applied worldwide if adaptable and profitable. The motivation for private research may be different from that for public-supported research. However, the outcome of both should be to enhance the capability and capacity for effectiveness of the total U.S. agriculture, food and fiber production, and delivery system.

Evolution and Response to Challenge by the Land-Grant System

In the U.S., the two major public entities supporting agricultural research are the USDA Agricultural Research Service and the land-grant university Agricultural Experiment Station system. Since much has been said recently about land-grant universities' agricultural research programs, it is appropriate to focus briefly on the historical as well as current challenges facing land-grant institutions. One additional reason for separating out the land-grant system is that much diversity of research is associated with the institutions that make up the land-grant system because of their they are located in different states.

The land-grant system was established to provide an educational program to develop human capital, generate new information through research, and extend research-based and validated information to the citizens of each state. Three federal acts--Morrill, Hatch, and Smith Lever--along with corresponding state legislation accepting the provisions of these federal acts, established the original framework for funding and developing the total land-grant system. Many subsequent federal and state laws and appropriations have further amplified the responsibility and capability for a member university or for the total system.

The Morrill Act was passed in 1862, the Hatch Act in 1887, and the Smith Lever Act in 1914. Congress, in approving the provisions contained in each act, established a concept of responsibility and actions for the state land-grant university. It was the belief of Congress that education (formal and informal) and research programs were essential to the solutions of America's problems. Further, they felt it was the responsibility of the federal government to invest directly in state educational and research programs for the overall good of the country. The concept of using the capabilities of a university to provide direct assistance to citizens was instrumental in creating the modern system of agriculture and quality education for many disciplines. The positive results from the U.S. land-grant system have been the envy of many countries. It is currently being studied in countries emerging from the breakup of the former Soviet Union. It is evident that Congress placed trust in the university system by giving them the responsibility for the development of new agricultural technology, application of technology, and generation of talent needed for this country to prosper and attain to its potential.

The land-grant system has undergone many changes in programs and educational and research emphasis as this country has matured. Currently, the system is not without strong criticism, internally and externally. Any system must invite both internal and external evaluation and respond positively to constructive criticism to remain strong. Programs of a land-grant university must be current and of value, both in reality and perception. It is recognized that individual universities may have responded differently to problems in their respective states and this variation in response should be expected. This variation is both a strength and a challenge for the total system. If a Congressional representative has positive views of his or her state teaching, research and extension programs, the support may be very good and this set of needs will at least be considered along with many other priority needs. However, if the representative's views are negative, or there is a lack of knowledge about the programs of his or her land-grant university, that support does not exist and other priorities are likely to prevail. It must be recognized that legislators have a myriad of problems presented to them. Thus, the land-grant system can expect nothing less than strong challenges in the economic environment of the nation and most states. There is tremendous competition for the public dollar in periods of great economic stress and unbalanced budgets. The land-grant system must present a forward-looking strategy in order to retain and/or increase funds. Further, the university must be adaptable and respond positively in periods of uncertainty.

A course of action must be defined to ensure that Colleges of Agriculture, Agricultural Experiment Stations, and Cooperative Extension Services maintain strong federal support. Even though this paper emphasizes research, the success of the total system is important to the success of research. There is the need for mutual support. The two entities are the primary recipients of federal formula funds within the land-grant university and will be the most influenced if federal support is lost or diminished. The base (formula) funds that come to a land-grant university primarily fund research and extension. Competitive and special grants are very important but it is base funds that permit the development of sustainable programs for research and extension as well as adoption of new programs. Below is a list of this author's recommendations.

* The land-grant system must present a unified front to decision makers.

* State differences must be solved locally rather than nationally.
* Decision makers must be convinced that the issues facing agriculture, rural America, the family, youth, and the community can be solved through education and research. Thus, it is important to aggregate needs, concerns, and problems in such a way as to be able to realistically demonstrate that education and research will help solve them.
* Research and extension programs must be designed to address current and projected critically important problems.
* Changes in programs must be communicated to both decision makers and system users and the programs maintained must address priority needs. Some efforts that have accomplished their objectives must be discontinued.
* The leadership of teaching, research, extension and international programs must cooperate and mutually support each other.
* Hard choices must be made and priority must be given to those programs that have a high payoff potential.
* The land-grant system must be continually revitalized and efforts made to communicate new directions.
* The confidence and trust in the land-grant system that has been lost must be regained.
* Land-grant universities must be viewed as agents driving change rather than resisting change. Land-grant universities must recapture the concept of changing their focus and programs as society and agricultural needs change. How one realistically communicates the expanse of responsibility of the total university in relation to the college of agriculture, Agricultural Experiment Station, and Cooperative Extension Service must be determined. This communication must show that the three entities have changed but should not have the total responsibility for change within the university system.
* Internal critics must take responsibility for system improvement. The criticism must be constructive and targeted.
* Land-grant leadership must be willing to address the concerns of both internal and external critics.

Current Status of Funding for Agricultural Research

The outcome of the environment for agriculture research in recent history has been a broad based erosion of support for public sector agricultural research. Base funds for publicly supported agricultural research have declined over many years in relationship to real purchasing power. It is also evident that public support of basic agricultural research has decreased in relationship to other programs and support areas in the federal scheme of funding (AAAS Report XVII Research and Development FY93). Agriculture is not the only area that has lost purchasing power in the basic research arena, but it has not enjoyed the gain that other federal agencies have experienced. Agricultural scientists have turned to other agencies and organizations for support of their research program, such as National Institutes of Health, National Science Foundation, and various private foundations, to compensate for loss of needed state and federal support. Support for agricultural research has actually been cut in a number of states. These cuts, when they occur, can be devastating to the research program, because a small cut may have a quantum impact on the research capacity of the Experiment Station.

The primary growth that has occurred in support for agricultural research in recent years has been in special and competitive grants. This has had a positive impact on opening new funds for research. However, the erosion in the purchasing power of base funds in support of many programs is critical for agriculture. This erosion may be more critical for agriculture than it is in some other areas of science because of the necessity to maintain the infrastructure of farms, field laboratories, animals, etc., that are required for agricultural production research. The amount of funds approved per grant project and the duration of support is less for agriculture than for other areas. More effort is required for equal purchasing power for agriculture than for biomedical research.

World competition in agricultural products has raised the question: should U.S. agriculture efforts be subsidized in order to compete in other subsidized markets? Within this context, some have viewed public sector agricultural research as a subsidy. The implication is a short-sighted view that emphasis on state-supported research should be decreased because a plentiful supply of products already exists. For example, the amount of wheat that is produced in Kansas far exceeds the market demand for wheat; therefore, some people question the continuing support of wheat research in Kansas. New varieties that are resistant to disease and insects would cease to appear if research were not continued, and it would require years to re-establish and organize research teams. Thus, it is critical to effectively communicate the long-term nature of research programs in addition to the short-term nature as exemplified by quantity of commodities that exist in the market.

From still other perspectives, world growth in agriculture research has been significant. Many countries have expanded their research programs to a relatively greater degree than the U.S. The percentage of world agricultural scientists located in the U.S. has been decreasing and the relative investment in research has been increasing in other countries. The implication is that U.S. agriculture is not as well supported research-wise as many individuals believe relative to other countries. Criticism of agricultural colleges has grown because some people and organizations perceive the college to be proponents of chemical agriculture. This criticism has the impact of raising questions on the need to support research if it produces detrimental environmental changes. It is clear that a number of organizations have an interest in reducing, not increasing, agricultural research or, at the very least, significantly shifting its emphasis. Finally, extreme competitiveness for the public dollar is occurring at both the federal and state levels. The demand for the public dollar exceeds the supply, and agricultural research is in competition with other agencies and organizations that are seeking funds. Supporters of agriculture research may not be as single-purposed in this goal as may be desirable to effectively compete in the appropriation process.

Public and private investment in research and development is very important to the total U.S. agricultural system. Even though total public investment in agricultural research has increased in the last 20 years, the investment has not kept pace with the increase in private research. Public and private investments were essentially the same in 1970, but are wide apart in 1990. Growth in research effort by industry should be viewed as positive for agriculture.

Future Actions

It must be realized that to answer the question where U.S. agriculture is going, the political environment that exists at both the state and national level must be evaluated. The key influences on future support for agricultural research and necessary actions can be highlighted as follows.

* Agricultural organizations and commodity groups are less powerful than they once were and are not as well organized as they need to be to effectively represent a minority base. Agricultural organizations must be willing to support research because they benefit directly from it. This support may not take place as readily as is desirable in periods when competition for the public dollar is great.
* Coalitions of agricultural groups have become extremely important. These coalitions must include not only agricultural organizations, but nonagriculture groups if they hope to influence the decision-making process in the U.S.
* Industry has certainly increased its efforts in targeted research. This is important and should be fully supported by public researchers.
* Central university administrators and governing boards have too often related the size of the agricultural population in the state or students enrolled in agriculture to the importance of agricultural research. This is brought about, in part, by the legislatures in each state trying to equate the population to the need. The political ramifications of this statement are understood, but it is not a correct comparison of the importance and need for agricultural research. The acres of crops grown or tons of product processed may be very similar to earlier times. There are definitely fewer people involved in agricultural production; yet it takes about 18 percent of the jobs in this country to produce and deliver food and fiber to the consumer.
* The complexity of agricultural problems appears to be much greater today than in the past. This requires a system and a cooperative approach to seeking solutions for many problems. The agricultural establishment is still learning how to develop good systems research.

Summary

Public sector research is absolutely essential to keep U.S. agriculture competitive in the future. The justifications are there for the development of new technology as a basic support for the U.S. agricultural industry. A real concern must be expressed whether the commitment by Congress and state legislators is sufficient to keep the community of public agricultural researchers strong and viable.

Finally, to answer the question *Where are we going?* It appears that the following picture is developing in this country. For the next several years, funding to support agriculture research will not be as great as at the present time due to the tremendous pressures upon state and federal resources. It appears that there will be a movement toward more targeting of resources to meet specific national needs. There is a likelihood that several land-grant institutions within the U.S. will significantly curtail research in agriculture, thus further reducing the capability and capacity of the total research institution in the U.S. While this may be a very solid forecast, it is a sad one because it foretells that a number of states may have little or no active agricultural research in the foreseeable future.

The U.S. is viewed as being a leader in agricultural research. However, we have the opportunity within the next decade to lose more of that leadership role than individuals perceive possible. When an action takes place, like reducing funding for agricultural research at the state level, and when it begins to happen in several states, the "snowball effect" can be tremendous in a system that is already overcommitted and underfunded relative to needs. As one analyzes the work with respect to the relative importance of basic and adaptive research, it must be emphasized that both are needed. There is real concern that we will not continue to have the funds to support both adaptive and basic research in the future. Adaptive research is required to make programs across the U.S. competitive and strong within the framework of a wide variety of environmental and other local conditions. The complexity of administration of agricultural programs will be greater as we try to fund efforts in the future. This complexity and increased need will be expected to be addressed with less funds rather than more funds. The priority-setting process within the U.S. agricultural research system must be honed to a high degree of effectiveness in order to allocate limited resources to areas of greatest payoff. This movement will have the effect of taking research out of certain areas that are important but in which the payoff period may not be as great. Thus, the recommendation is that the U.S. agricultural research community set appropriate priorities and not continue trying to do everything in light of the present limited resources. If one continues to try to support every area, the outcome will be one of reduced effectiveness and competitiveness rather than enhancement of the total program.

The strength of this position of U.S. agriculture is due in large part to the past significant accomplishments of the U.S. publicly supported research program. Without publicly supported research, the U.S. agricultural system would never have gained some of the international prominence that it enjoys today. The opportunity for continuing this pre-eminent role requires that the public continue to make significant investment in agricultural research. This investment has paid dividends not only in a wholesome and safe food supply, but at a very competitive price.

The perception is that there is a lot of technology in the pipeline that is ready to be adopted. No doubt this is true in some cases and with some producers; however, many of the problems we are facing today require a continuing effort to seek solutions. In the future, there will be a much shorter time between the development of new technology and its adoption by the producer. This places increased responsibility upon the publicly supported research establishment in this country, rather than less. It is critical that U.S. agriculture interests rise to the occasion and keep publicly supported research for agriculture strong and effective for the future.

Chapter 3

Agricultural Research in a Time of Change

Lawrence Busch
Michigan State University

One hundred years ago, the well-known soil scientist E. W. Hilgard was contemplating what Frederick Jackson Turner was to call the closing of the frontier. Hilgard (1882:651) argued that

> It is obvious, then, that so long as unexhausted soils and an abundance of 'fresh' land shall enable the cultivator to obtain...what he considers abundant returns, his interest in agricultural improvement and education will be but slight...."

In short, Hilgard realized that the closing of the land frontier would open another frontier, that of science, technology, and education.

With the advantage of 20-20 hindsight we can now all agree that Hilgard was right. Once it became apparent that there were limits on even the enormous abundance of land in the United States, farmers, farm journalists, bankers, and others became convinced that major changes had to be made in the practices used to care for the land and the crops and animals on it.

Today, we stand at a similar juncture. If Hilgard could write of the importance of the closing of the land frontier to agriculture, today we may write of the closing of another frontier. Its edges are still blurred and it will only become fully apparent to our children. But even now we may discern, as Hilgard did, its major features. For some time now farmers and their families have not been the majority of the population; today they are rapidly losing their special status (Browne et al., 1992). No longer are farmers universally viewed as the stewards of the environment; they are seen by some city folks as the spoilers of that environment. No longer are farmers seen as the rightful and sole determiners of farm policy; environmental and consumer groups are demanding their place at the legislative hearings as well and in some cases it is these latter groups that are now the major players.

Science, too, is changing. Scientists and policymakers are asking with increasing frequency about the post-harvest components of agriculture, about value added after the products leave the farm. The Land-Grant Universities, once almost exclusively the servants of farmers, are now besieged by a wide array of conflicting demands that far exceed their financial capabilities. And, the once sacrosanct philosophy of science that dominated agricultural research is changing as well.

These changes are with us now. They are not fleeting because the changes that rural America has witnessed over the last 50 years are essentially irreversible (barring a worldwide holocaust that would reduce us to iron age subsistence agriculture). Surely there will be surprises on the road ahead, but the changes are profound ones and our responses need to be equally profound. We need the wisdom of a Hilgard, a Bailey, a Davenport, to guide us through what are likely to be turbulent years ahead as a new form of agriculture begins to take shape.

In this paper I begin by describing some current trends that are having profound effects on agriculture. Then, I examine the implications of those trends for agricultural research. I conclude by asking whether we have the wisdom and the will to transcend our individual interests so as to serve the public good in the more complex era that is upon us.

Some Current Trends

Population. Perhaps the most obvious of all the current trends is that of population. There is little doubt that it continues to move inexorably upward, adding another Indonesia to the world each year. Yet, at the same time, it is overwhelmingly the poorest people in the poorest nations that have the highest birth rates. Thus, while the need for more food will continue to rise for the foreseeable future, many of those who need it will be unable to pay for it.

During the 1950s and 1960s the US could afford to have a food policy that involved the distribution of massive quantities of essentially free food to the poorest nations of the world. This occurred as a result of a unique congruence of domestic and international policy. American farmers were faced with massive surpluses that could not be consumed domestically and these could be distributed to the newly independent nations of the Third World as part of American foreign policy.

However, the oil crisis and the waning power of the US dollar combined to make this approach too costly by the late 1970s. As a result, free food distribution ended, leaving many developing nations with cheap food policies that they could no longer afford (Friedmann, 1982).

Today, despite continued hunger and malnutrition, the world economy has changed such that (1) many people in developing nations are now without access to land,[1] (2) effective demand for foodstuffs is unlikely to keep pace with population growth, and (3) many nations are no longer able to export food at subsidized prices without undermining their balance of payments. As a result, agricultural research in developed nations is unlikely to provide food for the developing world.

Environment. In the 1960s, the environment could be dismissed as a fad. Entomologists, when confronted with the evidence cited by Rachel Carson (1962), could argue that if there was a problem, it was surely the province of some other discipline (Dunlap, 1981). By the 1970s, however, it was no longer possible to ignore the forces of change affecting not only American society but the world as a whole. The environment has proved to be a significant source of social concern that refuses to go away.

Moreover, despite the myths of the farmer as steward of the land, environmental concern about the effects of farm practices is actually increasing and it is often met with hostility from the farm community. Farms are now the largest source of non-point source groundwater pollution in the US. Urban neighbors complain of the odors and effluents of large scale animal operations. Concerns about chemical pesticides persist.

Global Change. Another dimension related to the environment is the complex process of global climate change. While there is general agreement that the quantity of carbon dioxide and certain other chemicals in the atmosphere has risen in the last several decades, there is much less agreement over whether there is a significant chance of warming over the next several decades. Even if there is such a change in climate, at least one report suggests that the effects on US agriculture are likely to be relatively small (CAST, 1992). However, the CAST report is silent on the more complex issue of changes in variation across years in yields of agricultural commodities. Farmers have always lived with considerable annual variation in rainfall, temperature, and other climatological variables. There is some evidence that modern varieties increase annual variation at the same time as they increase yields. How global climate change would affect annual yields remains unclear. However, the development of varieties that can tolerate greater variation would appear a prudent research strategy.

Science Policy. The area of science policy may be divided into three discrete, though interrelated, issues: (1) waning support for science for science's sake, (2) the collapse of the philosophical underpinnings of reductionist science, and (3) the disjuncture between disciplinary problems and social issues. Let us examine each of them in turn.

Science for Science's Sake. In 1945, Vannevar Bush (1945) could well talk of science as the endless frontier. It appeared to all that the latter half of the twentieth century would be golden age of science and technology. Using the knowledge and wisdom generated by science, the world would be made a better place. More importantly, this would be accomplished by simply giving scientists more money for research. The federal government would simply become the world's largest supporter of science, dispensing thousands of grants to an ever-growing scientific community that would produce results that effortlessly would be transformed into technologies that would benefit all humankind.

One immediate effect of this new largesse was the dwarfing of agricultural research expenditures by those in other fields: health, the military, and even basic research soon came to make up the bulk of the federal science budget. Moreover, unlike the formula funding of agricultural research and its individual investigator focus, the new sciences were funded through competitive grants and often to teams of scientists. In physics, in particular, and in biology more recently, the age of big science (Price, 1963) had begun.

The bubble has now burst. No longer does the magic of science bring forth generous funding as it once did. As representative George Brown--arguably the strongest supporter of science in the Congress--recently argued at a meeting of the American Association for the Advancement of Science, merely pleading for more funds no longer is sufficient. Members of Congress see no reason whatever to favor scientists over other special interests: welfare mothers, farmers, highway builders, or the like.

Today, those who plead for science, including agricultural science, must make it clear that the science that is proposed will have clear benefits for at least some important segment of society and preferably for the society as a whole. This is at best a difficult task, although, as the funding for the National Research Initiative demonstrates, it is possible even in times of tight budgets.

However, there is a catch: Failure to deliver the benefits claimed can lead not only to discontinuance of the programs involved but to the reduction of the credibility of the proposers. The National Research Initiative may have within it the seeds of its own destruction as it is still very much supply-driven rather than demand-driven science (Busch, et al., 1991). What if the supposed beneficiaries reject the products generated by the molecular biology that occupies center stage in much of the NRI funding?

The New Philosophy of Science. Some thirty years ago Thomas Kuhn (1970) suggested that the building block theory of science was invalid. To Kuhn, science was better understood as going through alternative phases marked by puzzle-solving and conceptual rebuilding. Kuhn's work proved to be a Pandora's box that has led to the virtual collapse of the philosophy of science that was taken for granted by most scientists for nearly a century.

In brief, the focus has shifted from what scientists write to what scientists do and, in so changing, it has become apparent to all that much of the conventional wisdom about science was simply erroneous. The new perspective is perhaps best summed up by Bruno Latour (1987) who argues that "science is politics by other means." The point is that, far from being apolitical, science is always actively involved in politics, but its political activities occur in the extension of laboratory practices to the larger social world. The scientist who is unable to convince her or his colleagues of the validity of what has occurred in the lab is a failure, while the scientist whose work is extended everywhere is a success.[2]

The (political) power of science can be seen in the successes of the Green Revolution. The Green Revolution was a success not because scientists were able to develop high yielding varieties, but because they were also able to get farmers to change their farming practices--their social behavior--so as to make their farms look more and more like experimental fields. Similarly, the development of bST will only be a success if it changes both the way the hormone is produced and the behavior of farmers, processors, retailers, and consumers. In short, it is not that technical changes cause social changes, but that every technical change is a social change.

Of course, scientists can continue to go about their business within the confines of their labs and pay little or no attention to the larger social world, but they do so now at their peril. With the recognition of the political power of science comes the countervailing need for social responsibility.

Moreover, modern science is no longer distinct from technology; it is totally reliant on a wide range of technologies without which the phenomena it wishes to reveal would remain forever hidden (Idhe, 1990). Therefore, even if an individual scientist is utterly uninterested in the application of his or her results, their embodiment in technical apparatus make them available for use in industry. It is for this reason that we can talk of the role of science and technology in economic development and of the supply and demand for science (Busch et al., 1991).

This philosophical shift is being played out in every agronomy department across the nation. Once the placid setting in which like-minded scientists worked on whole plants with the central goal of increasing yields, traditional agronomists and breeders are being attacked on both sides. On the one hand, molecular biologists have entered the field and brought with them an entire toolkit for modifying plants at the cellular and molecular level (sometimes accompanied with some *naïveté* about how plants perform in the field). On the other hand, agroecologists have entered the field bringing a systems perspective previously missing. Each of these groups courts a different clientele and serves different societal interests. Whether they can find happiness in bed together remains to be seen.

The Disjuncture among Disciplines and Issues. Several hundred years ago, there was a close fit between disciplines and social issues. Those who studied mechanics could help in the design of more efficient steam engines; those who studied chemistry could develop new chemical compounds of interest to industry. To a great extent this is still the case. However, a new class of problems has arisen, which demand expertise that no longer fits within disciplinary boundaries. These problems are typified by the debates over global change described above. Their "resolution" will require new forms of interdisciplinary cooperation between atmospheric chemists and economists, between sociologists and agronomists to address issues that have fallen through the cracks between the disciplines.

In sum, the waning support for science for science's sake, the collapse of the received philosophy of science, and the disjuncture between disciplines and issues will have a profound effect on the ability of agricultural research to weather the next several decades.

Consumer concerns. Consumers are increasingly concerned about what they eat and how it is produced. While industry spokespersons and researchers have attempted to frame the debate around food safety issues--and these issues are of considerable importance to consumers--concerns are far broader than that. As I have argued elsewhere (Busch, 1991), consumers are concerned about food not merely as the fuel that keeps the human organism alive, but for the meanings that it conveys. That is why we offer our guests real butter and high fat, high sugar desserts, even though we may eat margarine and Nutrasweet™ at other times. It is why we consume different foods on holidays and other special occasions than we do on ordinary days. It is also why fried termites, orange sea slugs and other such delicacies are usually rejected by Americans despite their high nutritional value and safety.

Trade. Despite much rhetoric about free trade, it appears that the world is moving away from national trade areas toward regional trade areas. The North American Free Trade Agreement, the European Community, Mercosul, the Commonwealth of Independent States, and other proposed regional trading blocs are likely to change the rules of the game in ways that will only become clear over time. However, it is unlikely that anything remotely resembling global free trade in agricultural products will emerge.

Furthermore, while the collapse of the Soviet Union and the East Bloc is likely to lead to greater US export demands in the immediate future, higher agricultural production in those nations will undoubtedly be the medium to long term result. At the end of the last century the Ukraine was the breadbasket of Europe and Hungary was the center of the European grain milling industry. It is not unlikely that both will reemerge in the near future.

Farm programs. To date, myriad farm programs have managed to weather the storm. Even the GATT negotiations are held up by the various programs aimed at agricultural commodities. Nevertheless, it is doubtful that commodity based programs will continue to be politically viable much longer in the face of massive government debt, a declining farm lobby, an ideology that is less and less credible (Browne et el., 1992) and pressures to separate income support from commodity payments. The result will be considerable as farming is restructured to reflect a new political and economic reality. Dairy may well move south. Some croplands will probably revert to pasture. And, the particular mix of crops found in each state will doubtless change.

The changes in the tobacco program since its inception give one an indication of how research might be changed as a result of farm program changes. Initially, tobacco allotments were based on acreage and research began to emphasize how to produce the most on the least land. This rapidly recreated overproduction and led to poundage allotments. Since then, research has emphasized minimizing inputs required to obtain a given level of output. The end of farm programs is likely to have analogous consequences for research on other commodities.

Debt. The US national debt is growing at a rapid rate. As it grows it is likely to make the budget for agricultural programs and public agricultural research flat or declining. The consequences for agricultural research will be significant. In particular, scientists will receive less and less funding from their State Agricultural Experiment Stations and will move further towards responses to requests from other funding sources. Let us consider the implications for agricultural research.

Implications for Agricultural Research

The 1930s was the golden age of the general farm organization. The Farm Bureau, the National Farmers' Organization, the National Farmers' Union, the Grange, and other organizations that represented diversified family farm-based agriculture dominated the national scene. The 1970s was the age of the specialized commodity-based organization--one for each major and many minor commodities. These organizations eclipsed the general farm organizations and often squabbled among themselves. They turned debates about farm policy in general into debates about the details of programs designed to benefit a particular commodity.

However, in the 1980s, the environmentalists, consumer groups and others began to enter the debate. The dwindling numbers of farmers and the more urban character of the legislative process meant that farmers have had to share center stage on agricultural policy debates with these other organizations. By the end of this decade it is likely that farmers will only play a marginal role in formulating agricultural policy including agricultural research policy. Yet, who has begun to form the new coalitions with environmental and consumer groups? Who has begun to educate the non-farm public about the role of agriculture?

The recent failures of agricultural research should make us reflect. Food irradiation is virtually a dead issue in this country. bST is likely to proceed down the same path. Perhaps many of the transgenic plants now at the field testing stages will also die on the vine. There are two lessons to be learned from these failures:

(1) *Agricultural research is still largely supply-driven.* If we can make a new kind of widget, we go ahead and do so without ever asking anyone if they want one. The new biotechnologies have multiplied the potential range of products and processes to be developed without providing us with any guidance as to which among them to chose. We need to turn the entire research system around so that it is largely demand-driven. EMBRAPA, the Brazilian corporation for public agricultural research has recently realized this and, after drastic budget reductions, they have undergone a strategic planning process designed to turn that huge organization around (EMBRAPA, 1992). We would do well to take a closer look at their approach.

(2) *Agricultural production is based on the existence of long and complex commodity subsectors through which a given commodity passes on its way to being consumed.* Merely satisfying the needs of the immediate client in the subsector (e.g., farmers) is inadequate as other actors in the subsector often have the power to block technical changes that they see as detrimental to them.[3] Research providers, whether public or private, need to pay much greater attention to subsector organization when planning research. Why be the villain by attempting to supply something that is unwanted when one can be the hero instead?

Conclusions

There are other constituents out there waiting to support agricultural research, though the research they desire is hardly the "let's add another bushel" type of research common in the past. As noted above, environmental groups, consumer organizations and others are best regarded as allies rather than enemies. Moreover, rural bankers and Rural Electric Cooperatives are concerned about the future of rural America--a rural America that is crumbling before their eyes. They desperately need research that will bring them new jobs and more viable rural communities.

The move toward more and freer world trade will also have repercussions for research. We know little about how to serve foreign markets, how to develop new products for those markets, how to meet foreign standards (Hill, 1990). This will require both technical and socioeconomic research, to forecast demand, to create demand, and to meet the technical and social needs of foreign buyers.

We will also need to create the social, economic, and technical conditions that make it possible for our farmers to keep more of the value-added in our rural communities. This will require new opportunities for farmer cooperatives, new forms of credit for farmers, and new technologies that reverse the trend toward the appropriation and substitution of farm products by industrial equivalents (Goodman, Sorj, and Wilkinson, 1987). It will also demand the examination of new crops and processes (Jolliff, 1989; USDA, 1987; OTA, 1991).

In 1917, Kenyon Butterfield (1918 [1917]) President of Massachusetts Agricultural College, addressed his fellow colleagues in

Washington at the annual meetings of the Association of American Agricultural Colleges and Experiment Stations. He noted that,

> In our investigation we still stress too much the goal of increased productivity as our great task. We still have too much faith in knowledge of the physical and biological facts and principles as all sufficing. There should be searchings of heart as to our policies and programs. Are they adequate to the needs of the new epoch?

The epoch of which Butterfield talked is over. A new one is beginning. We would do well to ask the same questions that he did.

Endnotes

1. In a recent volume, Lipton and Longhurst (1989) have argued that the widespread problem of landlessness in Asia is likely to be replicated in Africa in the near future.

2. The literature on the new philosophy of science is already quite substantial. See Fuller (1988) and Rouse (1987) in particular.

3. An "economic" model of research and several examples of how actors can block technical changes that they see as not in their interests is described in Busch et al. (1991).

References

Browne, William P.; Jerry R. Skees; Louis E. Swanson; Paul B. Thompson; and Laurian J. Unnevehr. 1992. *Sacred Cows and Hot Potatoes: Agrarian Myths in Agricultural Policy*. Boulder, CO: Westview Press.

Bush, Vannevar. 1945. *Science: The Endless Frontier*. Washington, DC: Office of Scientific Research and Development.

Busch, Lawrence. 1991. "Biotechnology: Consumer Concerns About Risks and Values," *Food Technology*. 45(4): 96, 98, 100-101.

Busch, Lawrence; William B. Lacy; Jeffrey Burkhardt; and Laura R. Lacy. 1991. *Plants, Power, and Profit: Social, Economic, and Ethical Consequences of the New Biotechnologies*. Cambridge, MA: Basil Blackwell.

Butterfield, Kenyon. 1918 [1917]. "The Morrill Act Institutions and the New Epoch," Pp. 43-59 in *Proceedings of the 31st Convention of the Association of American Agricultural Colleges and Experiment Stations*. Washington, D.C.

Carson, Rachel. 1962. *Silent Spring*. Boston: Houghton Mifflin.

Council for Agricultural Science and Technology. 1992. *Preparing U.S. Agriculture for Global Climate Change*. Ames, IA: Council for Agricultural Science and Technology.

Dunlap, Thomas R. 1981. *DDT: Scientists, Citizens, and Public Policy*. Princeton: Princeton University Press.

Empresa Brasileira de Pesquisa Agropecuaria. 1992. *II Plano Diretor da EMBRAPA, 1993-1997*. Brasilia: EMBRAPA.

Friedmann, Harriett. 1982. "The Political Economy of Food: The Rise and Fall of the Postwar International Food Order," *American Journal of Sociology*. 88(Supp.): S248-S286.

Fuller, Steve. 1988. *Social Epistemology*. Bloomington: Indiana University Press.

Goodman, David; Bernado Sorj; and John Wilkinson. 1987. *From Farming to Biotechnology: A Theory of Agro-Industrial Development*. Oxford: Basil Blackwell.

Hill, Lowell. 1990. *Grain Grades and Standards: Historical Issues Shaping the Future*. Urbana: University of Illinois Press.

Idhe, Don. 1990. *Technology and the Lifeworld: From Garden to Earth*. Bloomington: Indiana University Press.

Jolliff, Gary D. 1989. "Strategic Planning for New Crop Development," *Journal of Production Agriculture* 2(January-March): 6-13.

Kuhn, Thomas. 1970. *The Structure of Scientific Revolutions*. Chicago: University of Chicago Press.

Latour, Bruno. 1987. *Science in Action: How to Follow Scientists and Engineers Through Society*. Milton Keynes, England: Open University Press.

Lipton, Michael and Richard Longhurst. 1989. *New Seeds and Poor People.* Baltimore: Johns Hopkins University Press.

Office of Technology Assessment. 1991. *Agricultural Commodities as Industrial Raw Materials*. Washington, D.C.: Office of Technology Assessment, US Congress.

Price, Derek J. de Solla. 1963. *Little Science, Big Science*. New York: Columbia University Press.

Rouse, Joseph. 1987. *Knowledge and Power.* Ithaca, NY: Cornell University Press.

United States Department of Agriculture. 1987. *New Farm and Forest Products: Responses to the Challenges and Opportunities Facing American Agriculture.* Washington, D.C.: New Farm and Forest Products Task Force, USDA.

Chapter 4

Agricultural Research -- the Role of Change

Richard A. Herrett
EnvirAg Associates

The subject of this chapter is the role and importance of change in agricultural research. Researchers have always lived with change. Biologists dealing with living systems have long since become acquainted with the notion of change and probably are least surprised by it when it occurs; whether its as a plant breeder attempting to develop a new variety resistant to stem rust or an entomologist that has just discovered a new insect strain resistant to a particular insecticide--change is common to all.

To illustrate one aspect of change, the model of the *Thomas W. Lawson* is useful to recall.[1] This was a seven masted sailing vessel built at the turn of the 20th century in response to the changes that were taking place in the shipping industry. Toward the end of the 19th century, steam powered vessels were taking away much of the business that the sailing industry had previously enjoyed and the seven masted *Thomas W. Lawson* was designed as a means to try and recoup that lost business. However, in the haste to regain lost business, the basic principles of sailing vessel design was overlooked and the story ended before it began. Before presenting the specifics of that story let us consider agricultural research and the notion of change.

The objectives of this chapter are: 1) to examine both the theoretical and real implications of change (innovation) on agriculture, 2) to consider the types of technology that have contributed to change, 3) to examine some of the factors that influence the evolution of changes now taking place, and 4) to examine the impact of change on the future of agriculture.

Theoretical Implications of Change

A S-curve (Figure 1) has often been a useful model to describe change whether it is in relation to the growth of a cell culture or increases in productivity. There are several important components of the S-curve. These include the initial lag phase which is followed by a rapid logarithmic increase, in the event is taking place, whether the number of cells or the production of a particular product. This is usually referred to as the growth phase. This phase is followed by a gradual slowdown to a plateau, sometimes referred to as the upper limit. The S-curve also serves to illustrate the theoretical impact when a single event such as a new technology is introduced in the absence of other competing technologies or events. In the real world, the upper limits, or plateau, of the S-curve for productivity in an industry has often been avoided as overlapping technologies emerge which can be aggregated to achieve ever increasing increases in productivity, even when many specific technology has reached its plateau.

The S-curve is useful for illustrating agricultural productivity as shown by the number of people supplied food and fiber - both domestic and abroad - by one agricultural worker (Figure 2). From the early 1800s until the 1940s, there was very little change in agricultural production, typical of the early lag phase demonstrating very slow growth. This was followed by a rapid rise in production from the 1950's into the 80s, typical of the logarithmic portion of the S-curve. The appearance of the plateau phase or the flattening portion of the S-curve has only become evident very recently.

Productivity more than doubled between 1940 and 1970 and slowed to increase at approximately of 2.2% per year between 1970 and 1985. This rapid increase in productivity enabled the U.S. to compete in the international markets and export food to other countries to the point where agriculture has become one of America's leading contributors to a positive balance of trade. The cumulative effect of this productivity can be demonstrated by the fact that today U.S. farmers produce 60 % of the world's food on only 13% of the world's cropland.

Technologies That Contributed to Change

It is certainly reasonable to conclude that this remarkable increase in productivity resulted from the input of many different technologies, or put another way, from the aggregation of many different S-curves. This can be demonstrated schematically (Figure 3) by displaying the various component technologies which contributed to this increase in productivity, including plant breeding, irrigation, mechanization, fertilizers and agrochemicals. Looking to the future, many believe that biotechnology is the next technology that will evolve and contribute perhaps even move significantly to productivity than resulted from the contributions of either the mechanical or chemical technologies.

This possibility raises an interesting question if in fact we are at the threshold of adding another S-curve with unprecedented capacity to enhance productivity, then why should there be concern about the future of agricultural research and the direction it seems to be heading? First, the potential contributions from biotechnology remain largely theoretical and untested. There is concern that the technology may be suffering a bit from some over selling brought about by the suggestion that it is "revolutionary" rather than "evolutionary" technology. It suffers some of the same problems that other new technologies have faced, a strong reluctance by the public to accept the products of that technology for a variety or reasons often times not related to the science driving the technology but rather to the public perception. Food irradiation is a classic example of a technology that has not reached its maximum potential.

Factors Influencing Change

Secondly, and perhaps having even more importance for the growth of productivity are the serious challenges facing agriculture. Some examples of these include:

* *The need to integrate national interests with global concerns*

* *Technological innovation that alters established techniques, processes and structures associated with the traditional food and agricultural enterprize system.*

* *Concerns about nutrition, food safety and climate change*

* *Fewer supporters of agricultural research resulting from the out-migration from rural to urban population centers. Such migration results in progressively less understanding of the frailties and vagaries associated with agricultural production and just how close we are to the knife-edge of starvation.*

* *Increased concerns over the impact of agriculture on the environment*

* *Less understanding of the issues compounded by the technical complexity of the science and the general reluctance of scientists to publicly support positions and state them in a positive way.*

* *Public discontent with science and demands for specific definable benefits*

* *Serious erosion of the agricultural research budget brought about in part by Congressional concerns about the USDA and its ability to meet the challenges of the 21st century.*

There is be no doubt that the past successes of agricultural research have come as a result of the traditional research, extension and teaching roles of the state university system. The system, however, has come under substantial criticism and appears to be headed for change which is both difficult to image and less than a happy thought for those within the system.

Changes began in the agrochemical industry in the early 1970's and took the industry from over thirty research-based organizations to fewer than 12 today. Names like Stauffer Chemical, Shell, Union Carbide and Lilly have disappeared either through acquisition or merger. The prospect of such changes are not ones any industry could look forward to, nonetheless they are essential to maintain competitiveness and thus to survive.

The emerging challenges outlined above are occurring at a time when there is both an increasing role in consumerism and an increasing public concern about the environment. These have resulted in major changes in the public affairs arena by traditional agricultural constituencies such as the agricultural commodity groups, processors, and producers. Nowhere are these pressures more evident than in the priority listings of the joint council on food and agriculture sciences--the organization established by the 1977 farm bill with responsibilities for establishing priorities for USDA research. To get an idea of the changes, Table 1 presents a side by side comparison of the priorities as established for 1989 with those recently established for 1994 as part of the budgeting process. The priorities are listed in the order of importance based on the 1989 list. Identical or similar topics in the 1994 are listed in a corresponding position.

Impact on the Future of Agriculture

A number of inferences about changes in priorities can be drawn from this comparison and provide insight concerning the future direction of research.

* Water retained its prime importance through concern for improved quality and limited quantities has been added.

* Biotechnology retained its position in second place, though there is additional emphasis on genome mapping.

* There was considerable expansion in the teaching area with the addition of more precise objectives focusing on both faculties and the curricula. There was also mention of increasing the quality and diversity of the work force in food and agricultural sciences and improving management strategies including adoption of innovative delivery systems.

* Improved understanding of the relationships between diet and health retained its rank in the two lists.

* The concept of enhanced productivity appeared in the 1989 list, though but only within the context of sustainability. The priorities in the 1994 listing which come closest to the notion of agricultural production include the adoption of environmentally sound management systems, promotion of safe and effective pest management strategies and the establishment of integrated plant and animal systems.

* The expansion of non-food uses for agricultural products appears on both lists. It is clear from the statements of former Secretary Madigan that he considers this to be an extremely important priority.

* In view of all the recent public concern, it is interesting to note that the issue of food safety appears as the last item on the priority list. The issue remains on the priority list for 1994 and incorporates an international dimension.

At a more general level, a comparison of the two sets of priorities indicates the following evolution of priorities:

* Increased productivity as a priority has evolved from an interest in productivity as to a position in which there is no mention of productivity in agriculture whatsoever.

* Sustainability is covered within the context broad interpretations of environmentally sound management systems and the establishment of IPM systems.

* Globalization has evolved as a priority with the listing of improved competitiveness in global markets through technology, education and policy.

* Social issues constitute the most significant differences between the two listings. These include reaching out to youth at risk, promotion of rural and community economic development, communication of the impacts of new agricultural technologies on people and communities, foster family economic well-being,

promotion of rural health and safety and the improvement of waste management in communities.

Certainly no one can argue that social concerns are unimportant. Perhaps what is of concern is the generalized shift from the concept of the importance of production agriculture and the incredible importance of new technology. It is appropriate to draw on the wisdom as espoused by Vern Ruttan in an editorial he wrote for *Science*. Dr. Ruttan stated:

> *"the capacity of American agriculture to expand its foreign markets depends on continued declines in the real costs of production. American agriculture has achieved its preeminence in the world by substituting knowledge for resources. This knowledge, embodied in more productive biological, chemical and mechanical technologies and the management skills of farm operators, has given the United States a world-class agricultural industry at a time when many other sectors of the economy are losing their preeminent position. A necessary condition for U.S. Agriculture to retain its status is enhancement of both public and private sector capacity for scientific research and technology development. The costs, to both consumers and producers of failure to maintain and enhance our efficiency in production would greatly exceed the adjustment costs resulting from abundance."*

In my view, Dr. Ruttan has stated very eloquently that we must continue to support those efforts which lead to new technologies. Within the context of Figure 1, only this will ensure continued expansion of productivity through continued emergence of technologies and associated increased productivity illustrated by the S-curve. It is through these new technologies that we as a nation can remain cost competitive and continue to expand sales of agricultural goods in foreign markets. It is especially important to remind ourselves of the importance of productivity when research budgets are declining and when there are so many other social aspects competing for limited resources. It is also important to remind ourselves of the direct correlation between the number of people required to provide food and the percentage of disposable income spent by the general public on food (i.e., The greater the number of farmers, the higher the percentage spent on food). It is clear the consumer has a real stake in productivity although often times this is overlooked in the emotional concerns over social and environmental issues.

Conclusions

In closing, let us return to the *Thomas W. Lawson* and the lesson we might learn. In the early 1800s, sailing vessels enjoyed a total monopoly for shipping cargo on the seven seas. There was however with the development of new technology, a gradual move toward the newer steam ships. As that trend increased, the sailing industry began to realize what was happening to them as they lost business; they had in the parlance of today's discussion, reached the top of their S-curve, in effect, they had reached their limit. In a last ditch effort to try and regain some of that lost business, they built a vessel they thought could compete with the faster steam ships and decided that a ship with seven masts could sail faster and thus be more competitive. In so doing, they overlooked one of the characteristics that made the sailing vessels so important-such as maneuverability under adverse conditions. When the ship was complete, they sailed it out into the harbor whereupon it capsized and sank in a mild wind and with it down went the shipping industry based on sailing vessels.

The lesson is not that American agriculture is about to go the way of the *Thomas W. Larson*, it is, instead, that we must not forget that the preeminence of American agriculture has been and will continue to be based upon its ability to train scientists and technologists and provide them with the resources necessary to make the basic discoveries that result in new technologies that lead to new S-curves. These skills can be combined with an extension capability that can disseminate the new technologies efficiently to those who have the skills to use them thus enabling the growth phases of the S-curve to continue with the addition of each new technology. We can ill afford to dilute the limited funding available to discover those new technologies necessary to continue the growth of American agriculture, for surely the capability which enables the American producer to compete with any producer anywhere in the world, will be diminished and we risk running aground because there are no new S-curves to be found.

Table 1
Priorities of the Joint Council on Food and Agricultural Sciences

Year	
1989	1994
Maintain and preserve water quality	Protect and enhance water quality and quantity
Expand biotechnology and its applications	Continue genome mapping and genetic enhancement
Develop and maintain scientific knowledge and expertise	Increase faculty competencies, including the use of emerging technologies
	Design curricula to meet scientific, technological and societal needs
	Increase the quality and diversity of the food and agricultural sciences work force
	Improve management strategies of the higher education systems including adoption of innovative delivery systems
Improve understanding of food, diet, human nutrition and health relationships	Enhance nutrition and food research and increase education on the relationships between diet and health
Sustain soil productivity	Adopt environmentally sound management systems
	Promote the safe and effective management of pests
	Establish integrated plant management and animal management systems
Assess new and expanded uses	Develop processes for new food and nonfood products
Preserve germplasm and genetically improve plants	(see priority 2 above)
Improve food processing, quality, distribution and for agricultural products safety	Ensure the safety and quality of food and agricultural products for domestic and international consumption

Figure 1

S-Curve

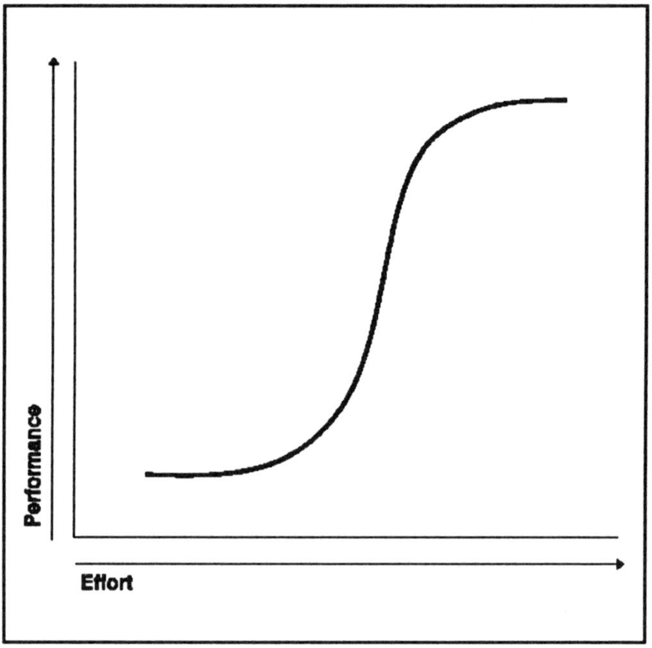

Figure 2

Agricultural Productivity as Shown by the Number of People Supplied Food and Fiber Per Farm Worker

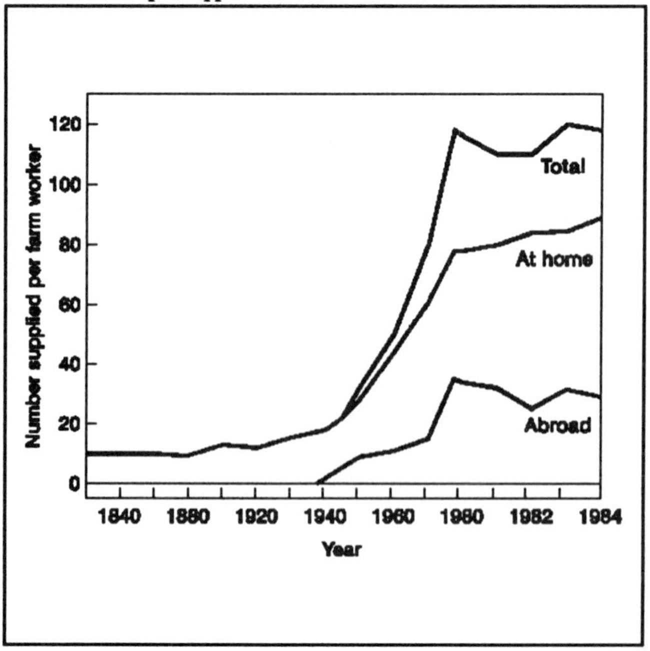

Figure 3

Impact of New Technologies on Productivity

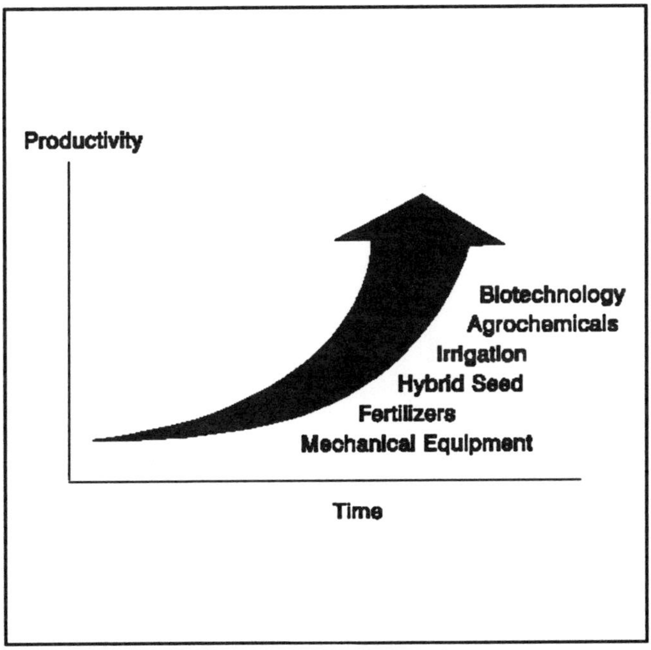

References

Foster, R. 1986. *Innovation, The Attacker's Advantage*. Summit Books, N.Y. pg. 28.

Herrett, R. A. 1988. Low-dosage Pesticides: Less is Better. *Chemtech* 18: 221.

USDA. 1992. Joint Council on Food and Agricultural Sciences, *1990 Accomplishments for Research, Extension*, and Higher Education, Preface, July.

USDA, 1992. Joint Council on Food and Agricultural Sciences, *Fiscal Year 1994 Priorities for Research, Extension, and Higher Education*, Executive Summary, June.

Ruttan, V. 1986. *Science*, Editorial, February 21.

The Current State of U.S. Agricultural Research: Empirical Evidence

Chapter 5

Trends in Food and Agricultural R&D: Signs of Declining Competitiveness?

Carl E. Pray[1]
Rutgers University

The National Science Board and leaders of the State Agricultural Experiment Stations (SAES) are concerned that declines in growth of U.S. R&D will reduce U.S. competitiveness in industry and agriculture. Many SAES and State Universities are suffering from declines in funding from state governments. State funding for the largest state agricultural research system, California, declined by five percent in 1990, four percent in 1991 and ten percent in 1992. In the industrial sector, U.S. nondefence R&D levelled off during the 1980s and early 1990s while our foreign competitors increased research expenditure according to a recent report by the National Science Board (June, 1992).

The main purpose of this paper is to present the most recent data on trends in public and private agricultural research expenditures in the U.S. and the rest of the world. In addition, the paper presents some preliminary analysis on how these trends affect the competitive position of U.S. food, agricultural, and agricultural inputs industries.

Public Funding of Research

The crisis of public funding of SAES research is a recent phenomena. The second line from the bottom of figure 1 shows that state funding of R&D in real terms grew from 1970 until 1990. The annual growth rate was 5.6 percent from 1971 to 1980 and 3.4 percent from 1981 to 1990. In FY1991 real state funding declined by one percent. The data are not yet available for 1992, but discussions with experiment station directors around the U.S. suggest that there will be another, probably greater, decline in 1992.

Figure 1 also shows that the U.S. government is the largest source of funds for public sector agricultural research in the U.S. The federal government provides about 53 percent of all funds to public agricultural research of which about one third goes to SAES and the rest to USDA agencies. State governments provide 35 percent. After a period of no growth from 1978 to 1989, federal funding grew sufficiently rapidly from 1989 to 1991 to more than compensate for the fall in state funds. In addition, funds from other sources such as private industry, commodity groups and sales of SAES products grew through 1991 so that they now account for 12 percent of public research funding. Thus, total public sector research increased from 1990 to 1991.

If the decline in state funds did not reduce total public agricultural research expenditures, did it reduce SAES expenditures? In real dollars, R&D expenditure by SAES did grow in FY 1991, but the growth was less than 0.5 percent. Furthermore, the slowdown in state funds came on top of a decade long decline in formula funds from the federal government. The combined decline in state and federal institutional support may have affected the allocation of research expenditures.

Two noticeable trends in research resource allocation are the increase in basic research using the new biotechnology techniques in the 1980s and the increasing share of production research from 1970s to the 1980s. The increase in biotechnology research is indicated by the number of faculty full time equivalents conducting biotech-related research at SAEs which increased from 273 in 1982 to 1131 in 1988. This increase took place at the expense of traditional plant and animal breeders and other more applied scientists (Hess 1991). The reduced importance of State funding and formula funding from the federal government has forced states to seek more resources from competitive grants distributed by the National Science Foundation, the National Institute of Health and USDA. These sources of funds primarily go to basic research with particular emphasis on biotechnology in recent years. Thus, the increased importance of biotech expenditure may be in part due to the changes in funding.

Research aimed at improving agricultural productivity has increased at the expense of environmental and demand-increasing research as shown in figure 2. The share of environmental research was clearly much greater in the 1970s than in the 1980s. It was not until the late 1980s that environmental research started to grow again. The increase in the importance of funds from industry, commodity groups, sales of farm products and royalties, and "other" sources of funds such as private gifts from 12 to 19 percent (see figure 3) may be partly responsible for the decline in environmental research. Other than private foundations, most of these groups are interested in production agriculture or utilization of agricultural products rather than the environment. Thus, their increased importance may have influenced spending away from the environment.

Private Research

The National Science Board recently reviewed private research and U.S. competitiveness in the economy as a whole (NSB 1992). They noted that after growing at an annual rate of 6.1 percent and 7.3 percent in 1975-80 and 1980-85, the growth of real nondefence research expenditure slowed to 1.6 percent in 1985-90 and is projected to grow 0.1 percent 1990-92. Although data on trends in private agricultural research is difficult to piece together,[2] available data suggests private research on food and agriculture has a different pattern. Research grew more slowly than all nondefence research from 1975 to 1985 (5.3 percent 1975-80 and 4.3 percent 1980-85) but more rapidly from 1985 to 1990 (2.6 percent).

Does this pattern simply reflect trends in sales or are other factors at work? Figure 4 shows that research intensity of the agricultural input industries increased in the late 1980s. Although it is difficult to see from the figure, research intensity in the food industry also increased from 0.18 percent in the late 1970s to 0.27 percent in the late 1980s. This suggests that something other than sales is causing the increase in R&D.

The growth in R&D and research intensity in recent years appears to be due to a combination of technological opportunities due to advances in molecular biology and stronger property rights. The advances in molecular biology gave researchers tools for conducting plant breeding and veterinary research more efficiently and made transgenic plants and animals possible. This was accompanied by court rulings which increased patent coverage to include microorganisms (1980), plants and plant parts (1985) and animals (1987). Figure 5 shows that the growth of agricultural research was led by the agricultural chemical industry, veterinary medicine industry, and the seed industry, which are all investing heavily in biotechnology, and by start-up biotech companies. By 1986, six percent of the research expenditure of the firms in the National Agricultural Chemicals Association was on biotechnology (NACA). In 1989, one third of the PhD scientists working in the seed industry were working on biotechnology (Kalton et al.). Farm input firms, such as the machinery industry, which are not based on biotechnology, declined from the late 1970s onward (Figure 5).

Total Research

Adding the public and private sectors together, the U.S. spends over $4 billion per annum on food and agricultural research. Figure 6 shows that total food and agricultural research has grown steadily through 1990 with no sign of the declines found in other industries. Private sector agricultural research has grown more rapidly than public sector research. It now accounts for at least 55 percent of all agricultural research in the U.S.

Competitiveness

The consensus of economists is that research stimulates the growth of agricultural productivity in the U.S. There is less agreement about the length and shape of the lag between research investments and productivity (Pardey 1986). Recent studies by Huffman and Evenson (1993) found that agricultural research conducted from 1 to 33 years earlier had a measurable impact on productivity. They assume that the impact grows gradually, and that research conducted from 8 to 13 years ago had the greatest impact with the impact tapering off in the years 14 through 33.

The relationship between productivity and international competitiveness is not as firmly established as the relationship between research and productivity because policies, the structure of industries, natural endowments and culture also affect competitiveness. Nevertheless, productivity does play a role in U.S. competitiveness and will play a larger role in the future as subsidies to agriculture and trade barriers decline.

In the period from 1961-65 to 1981-85, U.S. public sector agricultural research grew fairly rapidly but agricultural research in the rest of the world grew more rapidly. As Figure 7 shows the U.S. share of world public sector agricultural research declined from 26 percent in the early 1960s to 17 percent in the early 1980s. Its share of public research in more developed countries declined from 39 to 30 percent.

In the mid-1980s, the U.S. still had the largest amount of agricultural research, but was that level of U.S. expenditure largest relative to the size of the agricultural sector? The dark shaded part of the bars in Figure 8 show public R&D intensity. U.S. R&D intensity was greater than that in Less Developed Countries (LDCs) and Southern Europe, about the same as Northwest Europe, but lower than Canada, Japan and Australia. The greater intensity of the latter group of countries is compensated by the large amount of private research in the U.S. as the shaded parts of the bar show in Figure 8. In Australia, about one percent of agricultural research is conducted by the private sector (Kerin 1989), in Japan about one third of food and agricultural research is private (Japan 1991), while in Northwest Europe, countries vary from about 50 percent in the U.K. to about 30 percent in France. Data on Canadian R&D are not available and my guess for LDCs is less that 10 percent private.

Since the mid-1980's, U.S. public agricultural research grew more rapidly than public research in most other countries. Real research expenditure in selected countries is shown in figure 9. In United Kingdom and Canada R&D expenditure declined by over 20 percent since 1984. Declining public agricultural R&D also occurred in the Netherlands, Ireland,(OECD) most countries in Latin American and the Philippines. For example, Argentina recently cut the number of scientists at their national agricultural research institute almost in half and EMBRAPA's budget in Brazil has declined dramatically. Only a few countries experienced real growth in agricultural R&D. These include Japan, France and the U.S. which are on the graph and

Spain. As the graph shows the U.S. is growing faster than Japan and France since 1984.

Data on private sector agricultural research expenditure are not available for most countries. The incomplete data that does exist suggests that more private agricultural research is conducted in the U.S. than in any other country. However, there is insufficient data to tell whether the growth of U.S. private agricultural R&D is larger or smaller than growth in private research elsewhere.

Data on private R&D by agricultural input and food industries are not encouraging. U.S. patent data indicates that in the agricultural input industries foreign research is catching up. In the 1970s, the ratio of foreign to American patents intended for U.S. agriculture was .25. In the 1980s, this ratio increased to .45 which means that almost one of every three inventions intended to be used in agriculture the U.S. was invented abroad (Evenson 1991). In the food industry, the expenditure data in figure 10 suggest that Japan is catching up to the U.S. and France is also gaining.

How these trends will affect the competitiveness of U.S. farmers is not clear. The seven to thirteen year lag in the maximum impact of public R&D, coupled with the faster growth of foreign R&D in the 1960s and 1970s, suggests that foreign agricultural productivity will grow faster than U.S. productivity during the early 1990s. Faster growth in the U.S. during the 1980s, however, suggests that the U.S. productivity will grow faster in the first decade of the next century.

This projection is complicated by the role of the private sector. First, more U.S. R&D is conducted in the private sector than in most other countries and we know less about the relative growth of U.S. and foreign private research. Second, the lag between research and productivity growth in probably shorter in the private than in the public sector, but there is little empirical evidence on the length of the lag. Third, and perhaps most important, a large share of the research conducted by European and Japanese agricultural chemical firms, veterinary pharmaceutical firms, and farm machinery firms is targeted at U.S. farmers because they are the world's largest market for these products. Thus, any decline in growth rate of private sector research relative to foreign research is likely to have little if any impact on the productivity of American farmers.

Future R&D Trends and Policy Implications

Food and agricultural research in the U.S. have grown steadily in the U.S. through 1990. The decline in state funds for agricultural research is a very recent phenomena and, as of 1991, increased federal funding made up for the state decline. Any decline in private research is too recent to be indicated by the crude data available at present. Anecdotal evidence and data from some states indicates steep declines in state funding of public research and in R&D by some firms since 1990. Therefore, *the Clinton administration must carefully monitor R&D funding to determine whether corrective measure will be needed in the near future.*

The data presented in this paper do not contain any strong evidence that the U.S. R&D is growing slower than our competitors or that U.S. agriculture will lose its competitive advantage because of inadequate R&D. The contribution of public R&D to productivity growth may not be as great in the U.S. as elsewhere in the early 1990s. Productivity growth from public R&D should, however, increase relative to the rest of the world again from the mid-1990s to 2010 due to the increases in U.S. R&D during the 1980s. U.S. public sector R&D may be slowing down in the 1990s, but so is most of the rest of the world. Only a few countries such as France and Spain had faster R&D growth rates than the U.S. in the late 1980s., while Germany and Japan R&D appears to be growing at about the same rate as the U.S. *Data on private agricultural R&D in the rest of the world are needed to monitor the relative position of the U.S.*

In private research, the U.S. biotechnology industry has some advantages over the foreign competition: A strong public sector research base in biology, stronger protection of intellectual property, a less time consuming and costly regulatory process for testing transgenic living things and the largest market in the world for agricultural inputs. Thus, much of the private biotechnology research will continue to be in the U.S. for the near future. *Future policies must ensure that the U.S. maintains these advantages particularly through investments in public research and through regulatory policies that balance consumers concerns with the needs of industry.*

Even if the rest of the world rapidly increases private R&D, it is not clear that U.S. farmers would suffer. The competitiveness of American farmers is not determined by where the input company is owned or where the basic and strategic research is conducted but by where the technology is applicable. If technology adapted to U.S. conditions is available when needed, the U.S. can maintain a competitive edge. At present, the evidence suggests that large foreign firms are targeting much of their technology to the U.S. market. Thus, private research will continue to strengthen U.S. competitiveness into the near future.

Figure 1. Sources of Funding for Public Agricultural R&D

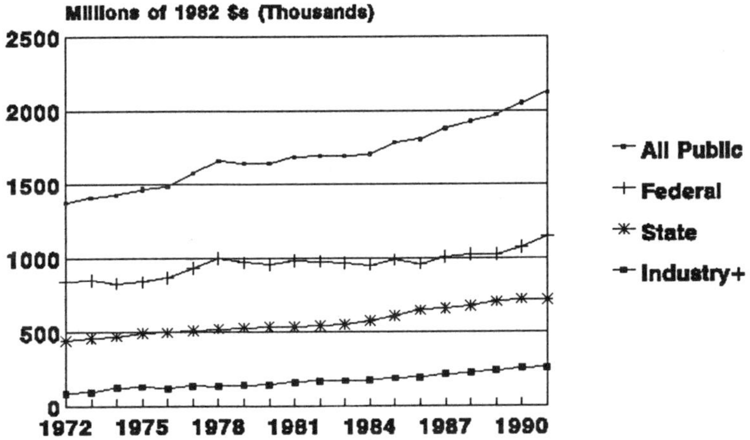

Source: USDA

Figure 2. Public Research by Goal

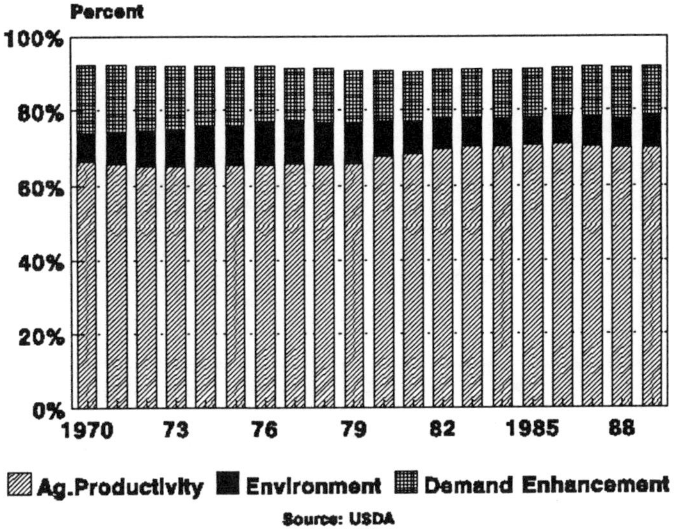

Source: USDA

Figure 3. Sources of Funding for SAES

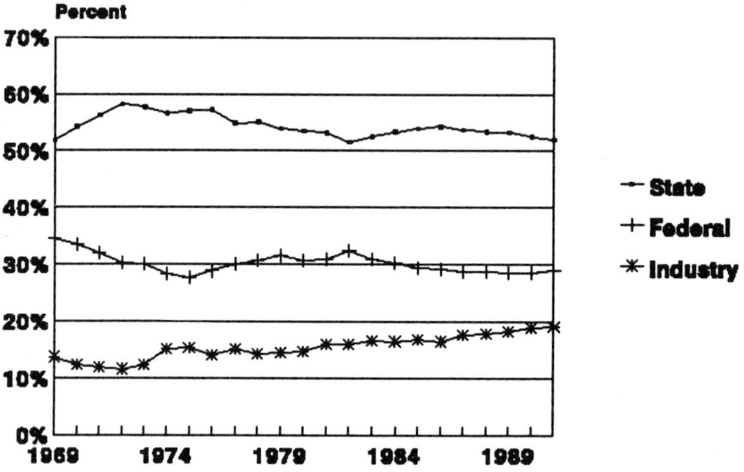

Source: USDA

Figure 4. Research Intensity by Industry

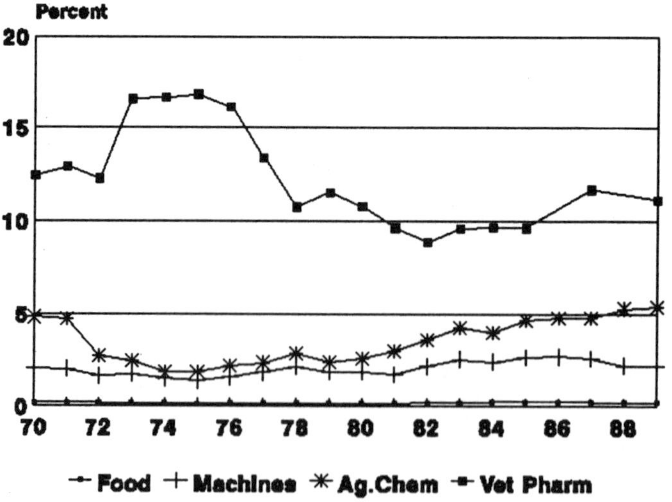

Sources: NSF, NACA, and PMA

Figure 5. Private R&D Expenditure on Food & Agricultural Inputs

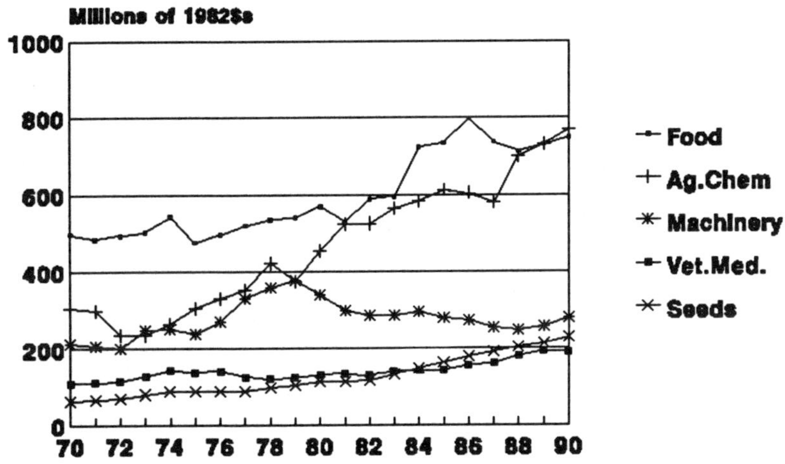

Sources: NSF,NACA,PMA,Pray&Neumeyer

Figure 6. Total U.S. Food & Agricultural R&D, Public and Private

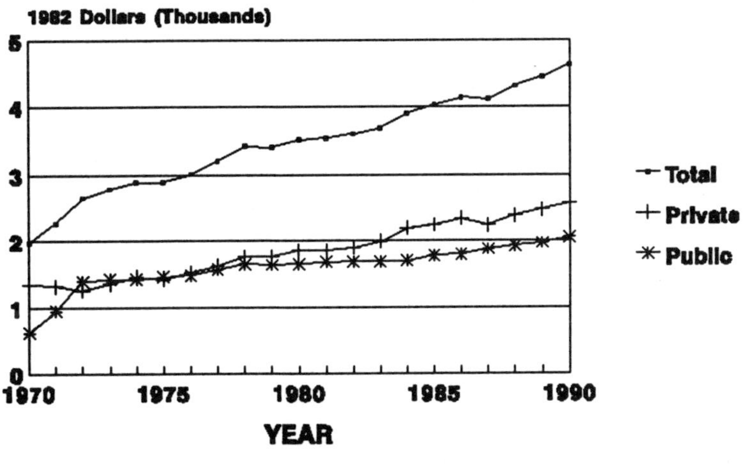

Sources:USDA,Pray&Neumeyer,NACA,NSF.

Figure 7: Agricultural Research Expenditures, Regional Shares

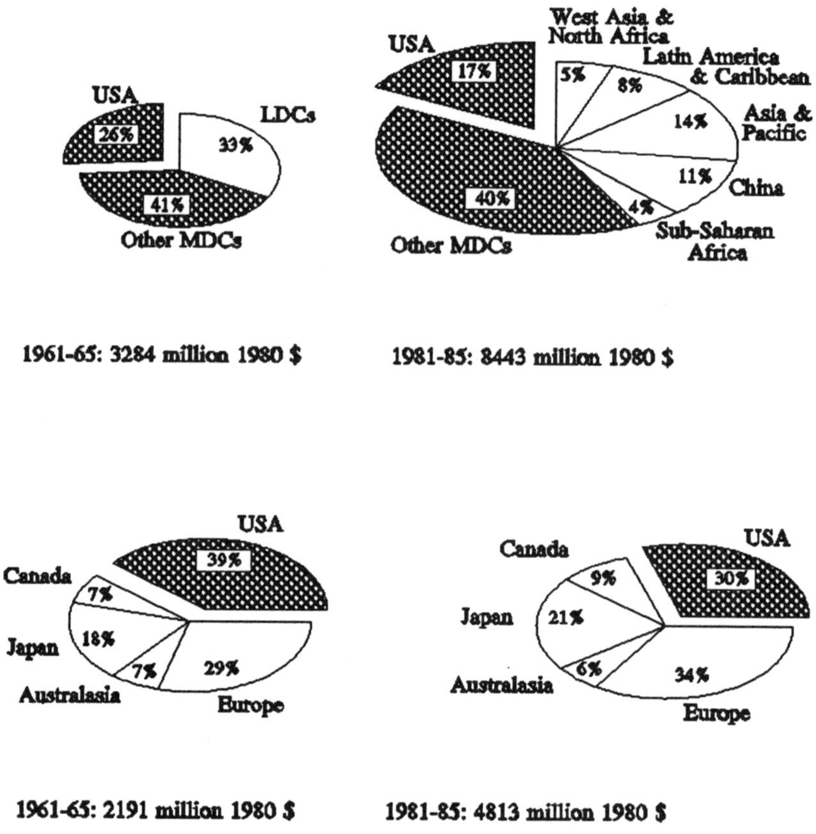

Source: Pardey, 1991.

Figure 8. R&D Intensity by Region & Selected Countries 1985

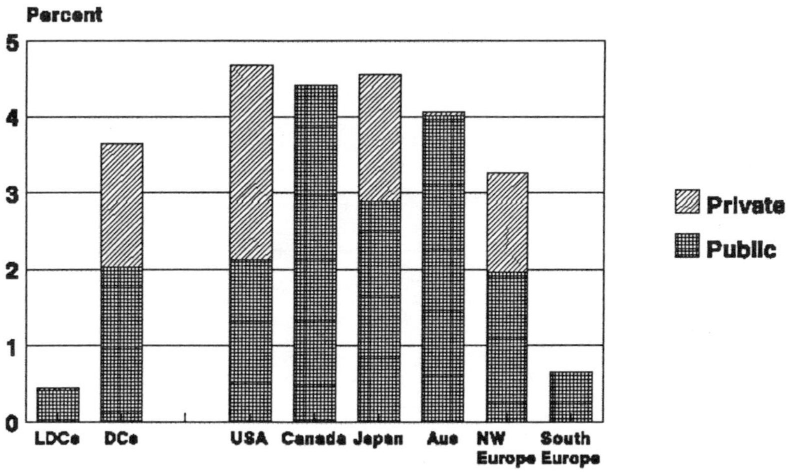

Source: Public R&D Pardey 1991, Private R&D Estimates of author (see text).

Table 9. Growth of Govt. R&D in Six Countries

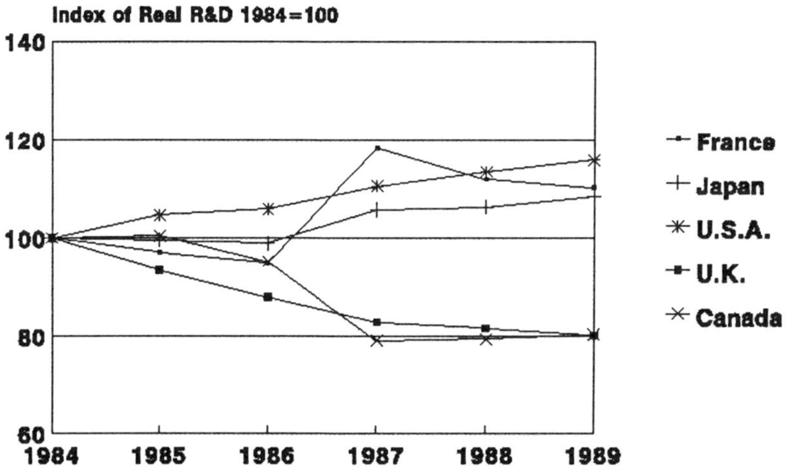

Source: OECD 1991

Figure 10. Private Food R&D in Six Countries

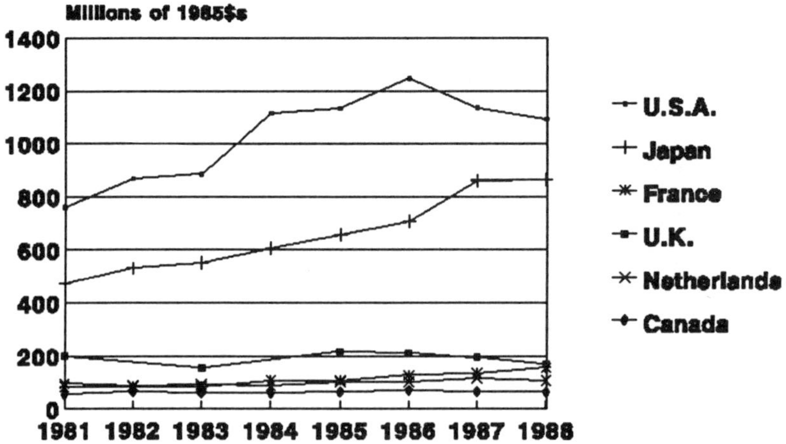

Source: OECD

Endnotes

1. I would like to thank Dennis Unglesbee of USDA for his assistance in obtaining U.S. public R&D data and Rob Weaver for his comments on a earlier version of this text.

2. The data presented here are based on NSF data and data from industry organizations and must be used with caution (see Appendix). No government organization surveys research conducted by the agricultural sector such as seed firms. Two agricultural inputs industries are three digit SIC code industries, agricultural chemicals and agricultural machinery. Research expenditure by these industries is available from NSF product field questions which are asked every other year and firms are not required to answer. Four digit industries such as veterinary pharmaceuticals are not surveyed at all.

References

Evenson, R. E. 1991. "Inventions Intended for Use in Agriculture and Related Industries: International Comparisons." *American Journal of Agricultural Economics*, LXXIII(3):887-891.

Kerin, John and Peter Cook. 1989. *Research, Innovation and Competitiveness: Policies for Reshaping Australia's Primary and Energy Portfolio Research and Development*, Australian Government: Canberra.

Hess, Charles. 1991. "Resource Allocation to State Agbiotech Research: 1982-1988." *Biotechnology*, IX(1).

Huffman, W. and R. E. Evenson. 1993. *Science for Agriculture*, Ames,IA: Iowa State University Press. Forthcoming.

Japan, Ministry of Agriculture, Forestry and Fisheries. 1991. *Development of Agriculture in Japan: Annual Report on Agriculture (excerpts) 1990.*

National Agricultural Chemicals Association. 1986. "Industry Profile Surveys," mimeos Washington DC: NACA.

National Science Board. 1992. *The Competitive Strength of U.S. Industrial Science and Technology: Strategic Issues*, Washington DC:National Science Foundation.

Organization for Economic Co-operation and Development. 1991. *Basic Science and Technology Statistics*, Paris: OECD.

Pardey, P. G. 1986. Public Sector Production of Agricultural Knowledge, Ph.D Thesis, Department of Agricultural and Applied Economics, U. of Minnesota. St. Paul.

Pardey, P. G. 1991. "Patterns of Investments in Agricultural Research: New U.S. and International Evidence." In *The Economic Impacts of Agricultural Research: A Workshop Sponsored by the Agricultural Research Institute*, Washington DC:Agricultural Research Institute.

Pray, Carl E. and Catherine Neumeyer. 1989. "Trends and Composition of Private Food and Agricultural R&D Expenditure in the United States," Rutgers U., Department of Agricultural Economics Research Paper P-0221-1-89. New Brunswick, N.J.

United State Department of Agriculture. 1970 through 1991. *Inventory of Agricultural Research*, Washington: USDA.

Appendix. Private Sector Research Data Problems

The basic problem of determining trends in private sector research is that there is no consistent source of R&D data for the agricultural sector, which includes the seed industry and farmers, or for the agricultural input industries. NSF collects data by product field every other year. These product fields include agricultural chemicals and farm machinery. However, in 1985 they stopped "imputing" the values of firms which did not answer the questionnaire that year and all product field series fell. The rectangles in Figure A.1 show what happened to the agricultural chemicals R&D series - nominal R&D declined by 26 percent between 1983 and 1985. Industry data, the triangles in A.1, showed a trend similar to NSF until 1983. Then they show 18 percent growth between 1983 and 1985. Unfortunately, the industry data then stops in 1987 when NACA started collecting data on members research worldwide which are shown by the pluses in A.1. In the latest NACA survey of their membership no R&D numbers were collected.

For other industries I have pieced together data from a variety of sources. Pray and Neumeyer (1989) describe these sources and the method of constructing the input industry R&D series in more detail. NSF is the main source for agricultural machinery although data on R&D by some of the large firms which is published in Business Week such as Deere & Co. also was considered. The veterinary pharmaceutical series is from the Pharmaceutical Manufacturers Association. The seed industry and biotechnology industry series were constructed using a variety of sources (see Pray and Neumeyer 1989).

The food industry series is based on the product field data from NSF. After 1983 NSF data on all R&D funded by the food industry was used to update the product field series.

Figure A.1. Agricultural Chemicals R&D from Different Sources

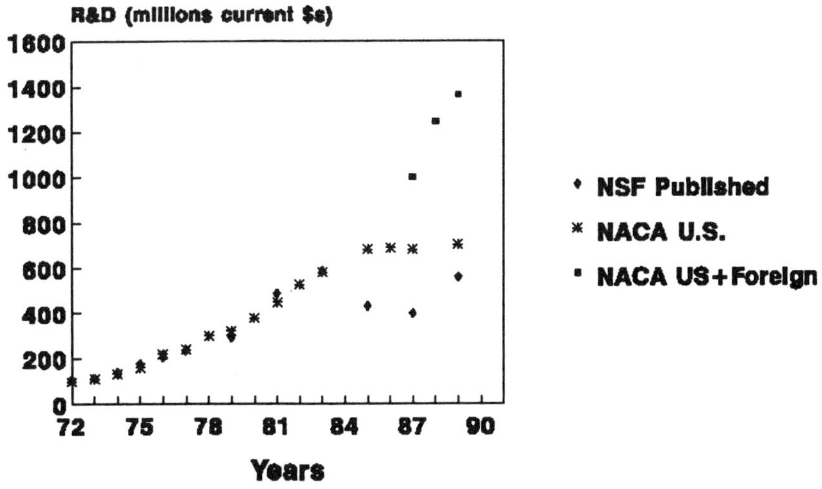

Sources: NACA and NSF

Chapter 6

Funding Agricultural Research: An Assessment of Current Innovation[1]

Donald N. Duvick
Iowa State University

When Neolithic grain farmers spread across northern Europe, moving slowly from east to west some 5,000 years ago, they invariably selected loess-derived soils for their farming operations (Bertrand, 1975). These easily-worked soils were suited to the simple plows of the early farmers, as well as being naturally productive for raising grain crops. Just how these early agriculturalists assayed the range of European soil types to find the ones they wanted is not known, but clearly they had an effective kind of agricultural research, able to develop and apply techniques of soil classification. How this agricultural research was funded is not known either. But since we are sure that their economy and social organization were simple, it is probable that "funding for agricultural research" consisted of finding enough extra food to support some inquisitive experimenters or an advance guard of prospectors charged with finding a new settlement site.

Decisions about funding, or non-funding, of research must have been easy -- no extra food, no research. But on the other hand, it may be that when food was scarce, applied research -- finding new farming sites -- was so important that "funding" went to the applied researchers at the expense of other, less essential members of the community. Today's agricultural research also is effective, and it also needs to be funded. I will discuss funding of U.S. agricultural research, dealing with past trends, sources of funds, uses of funds, and possible changes in future funding.

The Funding Record: Continual Growth

U.S. agricultural research funding has increased steadily since the earliest records in the early nineteenth century. This seems remarkable, in view of the continually declining farm population, now only 2 percent of the total U.S. population. During the past 100 years, real growth (growth in constant dollars) in funding for agricultural R&D has averaged a compounded rate of between 4 and 5 percent per year (Huffman and Evenson, 1993). Research and development (R&D) has been performed by both public and private organizations (government and industry), and the rates of growth of both kinds of R&D have been similar over the past century, although industry's rate of growth, on average, has been slightly higher (4.8 percent per year for industry R&D, versus 4.2 percent per year for public R&D). Private R&D expenditures have been greater than those for public R&D since about 1950. During the past decade private R&D has grown much more rapidly than public R&D -- 3.1 percent per year versus 0.4 percent per year, in constant dollars. Private R&D expenditures now are nearly double those for public R&D. In 1990, they were $3.2 billion versus $1.7 billion, in 1984 dollars (Huffman and Evenson, 1993).

Declining Rates of Growth

These last figures point out that while growth in funding remains positive, growth rates in agriculture R&D funding are now declining, especially in the public sector. Public R&D can be divided into that performed by the USDA and that performed by the state agricultural experiment stations (SAES). The growth rate in public funding for R&D in the USDA has not increased (in constant dollars) during the past 15 years, and has actually declined over the past decade (an average reduction of 1.1 percent per year during the period 1981-1990). Growth rates in SAES funding, in contrast, have increased consistently during the same period. But these have increased at a rate of only 1.0 percent per year compared to the long-term growth rate of all public sector funding of 4.2 percent per year. Thus, in the public sector, USDA funding is declining, and the growth rate for SAES funding is reduced compared to earlier years. In the private sector, the growth rate for R&D funding also has declined, but to a smaller degree. Its current rate of growth, 4.2 percent per year, is 0.6 percentage points per year (or about 13 percent) less than its long-time average rate of 4.8 percent per year, and 1.1 percentage points per year (or about 21 percent) less that its peak growth rate of 5.3 percent per year in 1945-1974.

In sum, according to Huffman and Evenson, during the past decade the change in the growth rate for USDA research funding has been negative, the growth rate for SAES funding is down by 40 percent, and private R&D funding is down 12 percent, in comparison to public and private long-term average rates of growth.

Relative Size and Comparison of Research Goals of Public and Private Funding Components

To put these differential rates in perspective one also should know the relative size of fund totals for the three categories of R&D: public-USDA, public-SAES, and private industry. During the past 40 years, the utilization of funds for all U.S. agricultural research has been (approximately) as follows: USDA, 15 percent; SAES, 25 percent; and private industry, 60 percent (Huffman, 1991). At present the proportion of private industry R&D is even higher, approximately 66 percent of the total (Huffman and Evenson, 1993).

However, to compare public R&D with private R&D is misleading unless one realizes that the two conglomerates generally perform different kinds of research. Further misconceptions can occur unless one knows that over time the several kinds of research performed by each conglomerate have varied in kind and proportion.

Today, private agricultural R&D can be subdivided into that which supports the food industry and that which supports production and sale of farming inputs: agricultural chemicals, farm machinery, and seeds. Most private R&D can be classified as applied research, using principles developed in basic research to develop new products and new uses for old products.

Public R&D more typically is of a basic nature, although product development -- applied research -- has always been an important component of public agricultural research. Public R&D

specializes in R&D innovations which the private sector will not pursue. The public sector's basic research typically is intended to develop new concepts or new procedures relating to agricultural practices and products. In general, applied research by the public sector is in development of needed products and concepts that are not suited to commercial development because of small market size (or even absence of a definable market), or poor ratio of development cost to affordable product price.

Funding Private Research and Development

Major changes in emphasis have occurred over the years, in R&D for the private sector. In the late nineteenth century support of the U.S. farm machinery industry dominated private R&D. This is not surprising, since at the turn of the century the farm machinery industry was the largest manufacturing industry in the United States (Huffman and Evenson, 1993).

Starting in 1945, private agricultural R&D began to increase emphasis on food products and additional agricultural inputs. Herbicides and hybrid corn are examples of new product lines that needed continuing applied research. Beginning about 15 years ago, biotechnology applied to agriculture began to rise in importance, and its share of agriculture R&D is still increasing. Agricultural chemical companies were spending about 6 percent of their research dollars on biotechnology, in 1986 (Pray, 1991). However, expenditures on biotechnology per se will be difficult to track in the future, as techniques of molecular biology become standard tools in various applied disciplines.

R&D funding for food products and for agricultural inputs both have risen during the past 20 years. Funding totals for both kinds of R&D are of similar magnitude, although funding for agricultural inputs R&D currently is about 25 percent higher than that for food products R&D. R&D funding (in real dollars) for food products increased very little from 1970 to 1980; it doubled in amount from 1980 to 1986, and it has since stayed constant in amount (Pray, 1991). R&D funding for agricultural input industries has varied markedly during the past 20 years, both between and within product lines (Pray, 1991). Funding for machinery R&D increased rapidly until the late 1970s, and since then has declined to about one-half its 1978 level. Funding for agricultural chemicals R&D increased steeply until about 1985 when it also began a steady decline, although the decrease to date is much less than that for machinery R&D. R&D funding levels for seeds and for biotechnology in agriculture have increased at a steady rate during the past 20 years. Funding for veterinary pharmaceuticals R&D has been more or less level during most of the 20-year period, but has shown a slight increase during the past 5 years. Of the four product categories (machinery, agricultural chemicals, seeds, and veterinary pharmaceuticals), seeds R&D has shown the strongest and steadiest growth in funding during the past 20 years.

At the present time, funding for R&D in agricultural chemicals is more than twice as large as that for any one of the other product lines (machinery, seeds, and veterinary pharmaceuticals). Those latter three categories are about equal to each other in amount of R&D funding at the present time. (But in 1970, funding for machinery R&D was double that of either seeds or veterinary products.)

Thus, over time the funding of private agricultural R&D has been at first more than, then less than, and now once again more than that for public R&D. Private R&D has varied over time, also, in degree of emphasis on different product lines and in nature of science used to do the work. But private industry consistently has emphasized applied and/or product-oriented research and it consistently has not duplicated public agricultural research without good cause to do so.

Funding Public Research and Development

USDA R&D. Public research funding also has varied through the years, in relative amounts among public entities and in emphasis on various kinds of research. Public funding was less than private funding in the nineteenth century but then surpassed private funding in the early years of the twentieth century, primarily due to large increases in funding of USDA research. In 1925, publicly funded R&D accounted for 68% of total agricultural R&D, nearly the reciprocal of today's public-private comparison. USDA R&D funding in 1925 was double that for SAES, again a mirror-image of today's comparative funding amounts for USDA and SAES. The period 1925-1932 represents a high point in funding of USDA research relative to other components of the agricultural research system (Huffman and Evenson, 1993). Public funding continued to exceed private funding until the

1950s, but then fell behind, largely due to sharp reductions in the post-war growth rate of USDA funding.

The rate of growth of USDA research funding has been erratic over the past century, whereas growth rates for SAES and private R&D funding have been more uniform. However, different segments and sources of SAES funding have varied in nature and amount over the years, as was shown earlier for industry R&D.

SAES R&D. Hatch Act appropriations started in 1888, but since they had no built-in provision for increases to keep up with inflation, they rapidly fell behind the needs of an expanding state experiment station system. (For example, Hatch Act appropriations were unchanged from 1894-1929, at $720,000 per year.) But other funding sources appeared, of four main types: 1) The USDA's Office of Experiment Stations, now the Cooperative States Research Service (CSRS), 2) Other federal government funds such as contracts and grants, 3) State government appropriations, and 4) Other non-federal sources such as commodity groups and private industry. These categories, while convenient, do not always contain what their names seem to imply, for through the years, classifications by government reporting agencies have not been consistent, across agencies or from year to year (Huffman and Evenson, 1993). They serve, nevertheless to paint in broad strokes relative amounts and changes in R&D funding for SAES.

Except for the early years, Hatch Act funds have provided only a small portion of total SAES funding. In 1955, Hatch Act funding provided only 4% of SAES funding for R&D. (1955 was the last year of non-amended Hatch Act funding.) Several acts of Congress provided for additions to the original Hatch Act funds, always giving all states equal amounts of money, as had the original Hatch Act. Then in 1935 the Bankhead-Jones Act provided significant additional funding and did so on a formula basis -- according to each state's share of the U.S. rural population. States also were required to match the Bankhead-Jones funds. Formula funding, a proportional system on a continuing basis, was now begun. It continued and rationalized the entitlements for SAES that had begun with the Hatch Act.

Additional acts of Congress added other sources of CSRS-administered funding, but the most remarkable change was the addition of the Competitive Grants program, in the 1981 Farm Bill. This program, also, is administered by CSRS. Although the funds in this program were small in total (and still are not large) they represented a dramatic shift away from the concept of entitlement -- of recurring distributions of formula funds to states for use by their land grant institutions. Researchers in all kinds of educational institutions, as well as private institutions and federal agencies, could compete for the grants. The grants were for a limited period of time and were not automatically renewed. Matching funds were not required. Most importantly, individual researchers were the unit of competition and of grant reception, whereas throughout all SAES history up to 1981, the state experiment station and its director had been the focus for CSRS-administered funding. The Competitive Grants Program was not an entitlement program.

Despite all the additions to CSRS-administered funds, their current contribution to SAES is relatively small, and has been growing smaller through the years. Thus, CSRS-administered funds fell from 25 percent of SAES funds in 1957 to 14 percent in 1990. The bulk of SAES funding (86 percent) comes from funds outside the control of CSRS. And most of that 86% comes from non-federal funds -- 74 percent of SAES funding came from non-federal sources in 1990 (Huffman and Evenson, 1993). State funds constitute a major part (73 percent) of the non-federal funds.

A proportional breakout of total fund sources for SAES in 1990 is as follows: Amended Hatch Act, 9 percent; Special Research Grants, 2 percent; Competitive Research Grants, 1 percent; Other USDA-administered Funds, 1 percent; USDA contracts, grants, cooperative agreements, 3 percent; Other Federal Funds, 9 percent; State Funds, 54 percent; Other Non-Federal Funds, 20 percent (Huffman and Evenson, 1993; Feltner, 1992). The first five categories in this list, rounding up to 14 percent of the total, are administered by CSRS; the others are not.

In sum, SAES R&D, although first funded primarily by the Hatch Act and other federal funds, now depends largely on non-federal sources for its funding. The states themselves provide the largest single share of funds, more than half of all SAES funds, on average. (However, states vary greatly, in proportion of support for R&D at their experiment station(s).) The second largest share of funding comes from other non-federal funds. Federal funds from all sources provide only about one-fourth of SAES funding.

Recapitulation: Agricultural Research Funding Up to the Present Time

During the past century, public and private sources have shared in supporting U.S. agricultural R&D. Relative proportions of the two sectors have varied over time, largely due to the erratic nature of funding for R&D to be performed by USDA. But individual components of SAES and industry funding totals also have fluctuated over time. Private R&D funds are now twice as large as those for public R&D.

Industry consistently has invested in applied research with the goal of increasing agricultural productivity. It concentrated first on machinery (starting in the 19th century), then on agricultural chemicals (beginning in about the 1950s), and now seems to be moving toward major emphasis on sophisticated improvements in food products, and in agricultural inputs such as seeds, and plant and animal protection. Industry, as individual companies, has supplied the funds for its own R&D, most of it performed in-house.

Public R&D in agriculture, consistently aimed at production of public goods through basic and applied research, has used a variety of funding sources, both public and private. The major effectors of public R&D have been the USDA and the state agricultural experiment stations (considered as a system although they operate independently). The SAES share of public R&D funds is more than twice as large as the USDA share. Individual states provide a major portion (54 percent) of SAES funds. In the past, most SAES funding has been of an entitlement nature and has been targeted to individual experiment stations, but there are signs that in the future important segments of funding will be non-entitled and will be targeted to individual researchers, competing with each other for funds.

Agricultural R&D funds, in total, have increased steadily through the years, at rates of 4% to 5% per year, but rates of growth now are decreasing, particularly in USDA funding, and to a lesser degree in SAES funding. (USDA funding has had a decade of flat or negative growth, and the rate of increase for SAES funding is declining. During the past decade the rate of increase in SAES funding was less than one-third that for the previous two decades: 1.0% versus 3.6%, per year.) Certain segments of industry R&D funding also are declining (machinery and perhaps agricultural chemicals), although the growth rate for industry funding as a whole is reduced only slightly, compared to its long-time average.

Future Trends in Funding U.S. Agricultural Research

Research and Development for Private Industry. Trends and percentages do not tell the whole story about funding of agricultural research, nor can they always give clear indication of future trends. In industry, more research expenditures does not imply greater productivity in product development. Particularly in the food industry and in research on agricultural chemicals like pesticides, increasing proportions of R&D now go into experiments designed to satisfy regulatory requirements (see Weaver, Huttner, and Townsend in this volume). Future regulations, aimed at giving risk-free food supplies or environmentally sound countrysides, can take profitability out of certain existing industry products but they also can add possibilities for new kinds of industrial activity.

Industry has a built-in governor for its R&D -- when a product line is unprofitable, or when a business consistently loses money, it is shut down. Conversely, entrepreneurial businesses continually try out potential new fields of commercial endeavor, including those in agriculture that require research for product development. Future directions in industry R&D will be different from present ones, just as present emphases are different from those of the past. But it seems impossible to predict, with any degree of certainty, what the new directions will be.

One safe prediction, however, is that industrial research in agriculture will be more sophisticated than in the past. Use of biotechnology will be part of the reason, but demands of the market-place for more sophisticated, better positioned products, and demands of the regulators for more refined data on product safety also will force researchers to place greater reliance on laboratories and electronic equipment, and to use the newest scientific advances in biology and the physical sciences.

Some observers speculate that government regulatory requirements will force small agribusiness companies out of business, and will hinder start-up of new entrepreneurial firms. This will happen, they say, because the small companies will not be able to afford the sophisticated testing and monitoring equipment that will be needed to satisfy future safety regulations. I think, however, that this will not happen. More likely, entrepreneurial testing services companies will appear, ready to serve the large market, in total, of small agribusiness firms. A new kind of private R&D will appear. (A few firms of this kind already are in operation. For example, private firms now perform molecular marker analysis on contract for seed companies. This information is used in breeding and selection, or as a technique for fingerprinting proprietary cultivars. As such technologies become more cost-effective, and as small companies learn more about them, the business of "high technology for hire" will grow.)

Industrial R&D for agriculture has changed its method of operation to a small degree in the past 10 to 15 years. Contract research with university researchers, or unrestricted grants to university researchers for conduct of needed basic research, have become more common.[2] Needs of biotechnology have been the chief impetus to this kind of off-site research. In general, it is used to supplement existing industrial research capacity, particularly when a new field of investigation is under consideration, or where a needed piece of research requires knowledge and equipment that probably would have only one-time use. As industry fills out its needs for staff and facilities to conduct biotechnology R&D, such off-site support may become less important, but it seems likely that a certain level of useful interaction will continue, on a greater scale than existed before the advent of biotechnology as a tool for agricultural research.

A certain amount of tension seems to be in the air, among those engaged in agribusiness R&D. Perhaps this is because change in direction and emphasis of research-based agribusiness companies is greater than in earlier years. Certainly the farm machinery industry is depressed and this must have affected both amount and kind of R&D in that industry. Manufacturers of agricultural chemicals, both fertilizers and pesticides, see greater regulation in their future and they are making long range plans with this in mind.

Pesticide firms, in particular, have been hedging their future by investing in plant breeding operations, hoping they can provide value-added plant protection (for a profit) by means of improved plant varieties. Often they (or their boards of directors) have believed that skill in biotechnology can give them a special advantage in developing products for the marketplace. Unfortunately, the newcomers in the seed business are finding that profit margins in seeds are not as high as they have been in pesticides, and that they are even negative for some seed product lines. They also are finding that help from biotechnology, while undeniably a future prospect, will not come in time to satisfy near-term or even medium-term needs for market-place advantage. Considerations like these probably have been factors in recent decisions of some companies to scale back their seed subsidiaries, or to sell them.

Because U.S. production agriculture is not particularly profitable at this time, and seems to have no large prospects for increased profitability in the near future, there must be a certain amount of pessimism among research-based agribusiness companies, at least those concerned with agricultural inputs. When farmers' profit margins are small and uncertain they are reluctant to invest in new technology or other inputs unless there is clear indication that the added expense will give substantial increases in profitability or stability of production. Although research can be used to advance one company over another (other things being equal), the industry as a whole cannot keep up its volume of business in a declining agricultural economy. To see more clearly what this means, one can contemplate the near-term and even medium-term prospects for profitability of agribusiness firms in the former Soviet Union.

But for some companies, times of change can be time of opportunity. As American agriculture changes, new opportunities for enterprising business firms will appear. However, how American agriculture will change is not readily apparent. It seems likely to some people that agriculture is moving toward greater integration up and down the food chain, and to larger farming operations. But another determined group of people believe it will be possible, instead, to move agriculture towards smaller farms with minimal dependence on agribusiness. They especially hope to reduce farmer dependence on products from large, non-local corporations. The former possibility indicates more work for industrial R&D, the latter possibility, by definition, leaves little room for industrial R&D.

In either case, it seems certain that American agriculture will depend more on biology and less on chemistry. This in itself seems to predict the future emphasis of R&D for agricultural business firms. Since agribusiness never has been particularly strong in biology with an ecological slant, new challenges surely await industrial R&D for agriculture.

Public Research and Development. Funding prospects for public agriculture research and development also will be affected by the profitability of agriculture, but they will be affected even more by the public's attitude toward agriculture, and how it weighs the importance of agricultural productivity as compared to agriculture's effect on the environment, rural society, and food safety. Current indications are that the balance will tip in favor of supporting research on the second group of concerns -- environmental and social concerns.

Such a turn of events probably would lead to even lower appropriations for USDA research, unless the USDA can convince the public that it (1) will abstain from helping to raise agriculture's productivity at the expense of the environment and rural society, and instead (2) will conduct research programs to give near-term improvements in rural environments and rural society, with only secondary emphasis on productivity.

Increases in federal funds for SAES production research probably will continue to slacken, for the same reasons. As a general rule, it seems likely that federal formula funds for SAES increasingly will be restricted in amount as time goes on, simply because of public perceptions at the national level that such funds are used only for production research that benefits the wealthy and harms the environment.

Prospects for state funding of SAES are not bright, either. The national economic situation seems to be affecting all state budgets in the same negative manner, with states differing only in the degree to which they are suffering from budget deficits and spending cutbacks. Routine appropriations for agricultural research will be reduced in all states, and in highly urban states they may be cut more than other sections of the state budget.

Enterprising SAES directors, and department deans, will look to other sources of funding, tailoring their research operations to suit donors' (or investors') desires. In general, SAES research will be pointed toward four goals, most of them relatively new: sustainable agriculture, rural sociology, biotechnology for agriculture, and product utilization. SAES directors will attempt to play to both sides of the house: to environmentalists, and to agribusiness (which I here define as including commercial farmers). Funding to support research in these fields will come from a variety of sources, and fund donors for environmental and sociological research are not likely to be the same as those supporting biotechnology or product utilization.

As with agribusiness R&D, the only sure prediction about sources of funding for public R&D in the future is that the sources will not be quantified in the same rank order as at present, and probably there will be some new sources, as well. I can speculate about future sources for public R&D, however, and will do so.

At present, 20 percent of SAES funds comes from an assortment of non-governmental sources such as: foundations, private corporations, farm and trade associations, commodity groups, and fees for services and sales of research products or by-products (Huffman and Evenson, 1993). This conglomeration of donors and sources may give the most promise for adding to SAES funds in the future, with opportunity to tailor the donor list to the research goals of the station.

For example, farm and trade associations and commodity groups will be likely to support well-described research in plant and animal protection, or in ways to increase productivity and profitability of certain commodities. These groups have long-term interest in such research, and it would be well for them to become accustomed to providing substantial and continuing research support. If public attitudes continue in the direction they now are trending, parties with direct interest in outcome of production and protection research are going to have to provide a significant portion of the funds in support of that research, or it may not be done. Similar advice could be given to agribusiness firms, regarding support of public basic research that specifically undergirds their applied R&D. Businesses have a collective, long-range self-interest in supporting certain kinds of public-goods SAES research. Of course as noted earlier, individual agricultural business endeavors (including farming) must be profitable, or they will be unable to afford such research support.

Foundations with interest in promoting environmental improvement conceivably could be educated about and become enthusiastic supporters of agricultural research intended to help agricultural producers (farmers) as they try to combine the sometimes opposing goals of environmentally-friendly farming and profitable farming. Other foundations would have more interest in funding agricultural research of a sociological bent, aimed at improvement of rural society. They, too, conceivably could be converted, to support SAES research that would further their special goals. Since foundation giving tends to follow bandwagons, such donor support might be tailored to SAES start-up programs, with intention to find longer-range support once superior new research programs were under way.

Foundations, like business interests, have long-term and perhaps even moral reasons for supporting SAES research that supports their goals. But the foundations' extreme lack of knowledge of agriculture (in most cases), or their conviction (in many cases) that SAES research is incorrigibly irrelevant and misdirected, will make it very difficult to prevail upon them to support SAES research. I believe one of the greatest needs of experiment station directors and college deans is to become part of the foundation establishment, to learn how to empathize with foundation directors, and eventually to persuade them of the utility of supporting certain kinds of SAES research.

State funding will continue to be the backbone of SAES support, despite current budget crises. SAES leaders already know how to deal with their individual state funding agencies and officials. But renewed and broader presentations at the grass-roots level may revitalize state legislators and state officials, as has been shown recently in Indiana. SAES leaders also may need to improve their acquaintance with some of the emerging new power groups whose interests in environment and society necessarily will have ripple effects on production agriculture and the rural countryside. Experience in Iowa has shown that such acquaintance can promote beneficial interactions, helping all concerned groups.

A great deal of concern is expressed on land-grant campuses about the effects on agricultural research of the Competitive Grants Program, and of other competitive grant opportunities, as well. A chief concern is that much agricultural research is long-term in nature with little visible outcome for many years, and so is best suited for support with recurring formula funds. Three-year bites are not suited to a plant breeding program with 15 year cycles, for example. Researchers spend inordinate amounts of time writing grant requests, with very low rates of success. Further concern is expressed, that when individual researchers are the unit of funding decision, administrators cannot mold overall research programs at a station.

Despite the above problems, I am not pessimistic about the effect of the Competitive Grants Program on SAES research. In the first place, since at present the funds are only 1 percent of the total SAES expenditures, they cannot have large effect on SAES performance, unless they have remarkably large leverage. (But it also should be noted that about 6% of state university funding now comes from competitive grants from several federal sources, and another 3% comes from federal contracts and cooperative agreements; Pray, 1991, Huffman and Evenson, 1993.) As with any new technology, efficiency in writing grant applications, and in getting grants, will improve with time. Some universities now offer courses in grant writing, I am told.

More importantly, the Competitive Grants Fund is modeled after a granting system that successfully has supported basic research in biology, medical research, and the physical sciences for nearly 50 years. Increasing numbers of SAES research programs are of the same general mold as those in biology and the physical sciences, and so should be as well suited as they, for support with competitive grants. (But disciplines may differ, in the degree to which they can efficiently use competitive grants.)

Finally, since researchers almost always do research that is tailored to their training and inclination, I believe directors can best mold overall programs by their hiring policies, not by trying to influence funding mechanisms.

One other novel funding device needs discussion: licenses and royalties. Land grant universities, and the USDA, have discovered intellectual property rights (IPR), applied to agricultural products and processes. Applications of biotechnology and plant breeding have been the special stimulus to use of IPR. Two factors are considered: use of IPR to allow a public institution to prevent undeserved or improper monopolization of a university's invention or new variety, and use of IPR to make money, via royalties.

The first option is laudable and sometimes may be in the public interest. A patented invention can be released to the public, thus fulfilling the SAES mission of producing public goods. This option also can be used to encourage marketing of an invention or variety with such a small potential market that it can be used by only one firm at a time. In such cases, open bidding could be used to select a firm that can be trusted to put the invention or variety, suitably patented or protected, before the public. This procedure would not raise much money for SAES R&D, for such occasions are rare. It simply would be a public service.

The second option -- patents as money-makers -- brings problems. Although there seems to be no reason, ethically, why universities and their employees should not patent their inventions or protect their new varieties, the reasons for doing so, and the ways in which royalties and license fees are distributed, can be disruptive of SAES and university goals. Despite much evidence to the contrary, university administrators often believe patents and plant variety protection certificates can be important money-makers for the university or the experiment station. The facts are that very few universities have made significant profits from IPR (when bookkeeping is realistic), and when they have done so, it has been on a single product or product line. Such lack of diversity is a sure invitation to financial failure, sooner or later.

As many boutique biotechnology companies could have said before they disappeared, patenting is not a business. Patents can help support a business, particularly a research-based business. If universities and experiment stations should become business firms, with product lines, production and sales departments, and marketing directors, they would find patents to be an invaluable aid to the business. Products, production, and sales make a business; patents help secure the business. To repeat, it is not unethical (in my opinion) for SAES to make money from their inventions and new varieties, but it is not wise to expect that such income will be either steady or substantial.

A further problem with royalty collection by the universities is the practice in many universities of dealing researchers in on a portion of the royalties of inventions they made while working for the university. Any business can say that this is a sure way to disrupt and hinder cooperation among the broad and continually changing groups of scientists now needed to bring forth any scientific advance. When employee contributions belong to the employer, the institution can be operated in the most productive manner. One must realize, however, that scientists who have solicited their own research money from outside the institution may feel that their allegiance is not entirely to the institution that pays their salary. This is a problem that needs solving.

In summary, SAES funding will continue to come primarily from the states, but funding by special interest groups, both commercial and public interest, increasingly will be solicited. Such solicitation will come both at the experiment station level and at the researcher level. SAES directors and researchers will have a special problem, to avoid appearance or actuality of being controlled by those from whom they solicit funds (either commercial firms or public interest organizations). Formula funding will continue but will represent a smaller portion of the total, as individual researchers become entrepreneurs supporting their own research through competitive grants. SAES administrators will find that their control of station research will need to be more indirect. Their direction, in the future, will be most effective at the policy level.

Summary and Conclusions

Public and private sources have shared in support of agricultural research and development over the past century. USDA, SAES and private firms have been the loci of agricultural R&D. R&D performed by USDA and SAES has been funded primarily by the public sector. Private firms have funded their own R&D. Relative proportions of funding, comparing public with private, or within any of the three major groups, have varied through the years. At present, funding of private R&D is double the amount for USDA and SAES combined, and funding for SAES is more than double the amount for USDA. States provide the largest share of SAES funding, other non-federal funds furnish the next largest share, and federal funds administered by CSRS furnish the third largest share of SAES funding.

Agriculture research funding, in constant dollars, has increased continually and substantially over the past century. The rate of increase now shows signs of slackening, primarily due to cut-backs in funds for USDA R&D, reduced growth rate of SAES funding, and reductions in funding of some kinds of industrial R&D. Imbalances in the national and state economies, poor profitability in agriculture, and public disinterest in support of production agriculture are causes of the reduction in funding support for agricultural R&D.

Future funding for private R&D will depend on profitability of research dependent agribusiness firms. The subject matter of agribusiness R&D will vary according to new kinds of products and markets. The nature of products and markets will be influenced by government regulations as well as by customer demand.

Future funding for public R&D will continue to be based on commodities and (for SAES) formula funding, but to a lesser degree. For both USDA and SAES, funding increasingly will be intended to support agricultural research that promotes environmental and social well-being in the countryside. SAES will continue to depend on the states for a substantial portion of funds, their direct federal support will continue to decline, and increasing proportions of fund totals will come from special interest groups (both commercial and public interest), and also from competitive grants to individual researchers. SAES thus will have a new and more diverse customer base, and will be more responsive to their customer base. SAES will have a special problem, to remain independent of their donors, as they continue to produce public goods.

Endnotes

1. For presentation to the Agricultural Research Institute's Symposium on the Dynamics of the U.S. Agricultural Research System, September 17-18, 1992, McLean, VA.

2. In 1969, funding from industry grants and commodity groups supported 5 percent of SAES expenditures. In 1989, comparable support equaled 7 percent of SAES expenditures (Pray, 1991). But industry support for biotechnology research in land grant universities probably is considerably greater than 7 percent of biotechnology budget totals (Buttell and Curry, 1991).

References

Buttel, F. H. and J. Curry. 1991. Relationships Between Industry and Land Grant Universities: A Survey of Agricultural Scientists. In, *The Economic Impacts of Agricultural Research*, A Workshop Sponsored by the Agricultural Research Institute, Washington D.C.

Feltner, K. C. 1992. Personal Communication, CRIS Data.

Huffman, W. E. 1991. Research-Productivity Results at the State and Regional Levels. In, *The Economic Impacts of Agricultural Research*, A Workshop Sponsored by the Agricultural Research Institute, Washington D.C.

Huffman, W. E. and R. E. Evenson. 1993. *Science for Agriculture*, Chapter 4. Forthcoming, Iowa State University Press.

Pray, C. E. 1991. Trends in R & D Productivity and Expenditures and Implications for State Agricultural Experiment Stations. In, *The Economic Impacts of Agricultural Research*, A Workshop Sponsored by the Agricultural Research Institute, Washington D.C.

Chapter 7

Trends and Issues in Agricultural Research Funding at the State Agricultural Experiment Stations

David B. Schweikhardt and John F. Whims
Michigan State University

The past two decades have witnessed important changes in the funding of agricultural research in the United States. Agricultural research funding at the state agricultural experiment stations grew rapidly between 1945 and 1970, with federal funds for experiment station research increasing by 10% annually in real terms (Office of Technology Assessment, 1981, p. 202). While the federal government provided a large share of the resources at the state agricultural experiment stations, these funds were allocated to the states as formula funds and managed by experiment station directors in response to demands from local research users. Recent years have seen a reversal of these trends, with growth in research funding having declined since 1970, particularly at the federal level. At the same time, the mechanisms used to allocate research resources have grown more varied in recent years.

The U.S. agricultural research system now faces a number of major policy issues related to the level of research funding and the mechanisms used to allocate research resources. This paper reviews recent trends in agricultural research funding at the state agricultural experiment stations. These trends are then linked to the underlying policy issues reflected by these funding trends. These underlying policy issues will be identified, and public finance theory will be used to draw some conclusions that are relevant to these issues.

The Level and Sources of Funding[1]

The real rate of growth in agricultural research funding at the state agricultural experiment stations has slowed in recent years. Real expenditures for research grew by 1.6% per year between 1979 and 1990, compared to a real growth rate of 5.5% per year during 1970 to 1979 (Figure 1). The sources of funding available to experiment stations have changed very little during the past 20 years (Figure 2). State appropriations constituted roughly 60% of total expenditures throughout this period. Federal funds from the USDA declined from 22% of total expenditures in the early 1970's to about 18% in recent years. A more significant change -- a change in how USDA funds have been allocated to the experiment stations -- will be discussed in the next section. Federal funds from sources other than USDA remain steady at 10% of total expenditures during this period, and industry funds have increased from 5% in the early 1970's to 7 or 8% in recent years.

Total funding per scientist has grown at 0.7% per year in real terms since 1979 (Figure 3). Federal funding from USDA-CSRS sources has remained at $25,000 per scientist for the past 20 years. Funding from state appropriations has experienced a small amount of growth since 1982, but this growth has only managed to return the level of funding to its 1979 level.

Mechanisms for Allocating Federal Agricultural Research Funds

Agricultural research funds are allocated from the USDA to the state agricultural experiment stations by way of three allocation mechanisms:

> *Formula funded grants* allocate research funds to the states based on a legislatively-mandated formula. The existing formula for allocating Hatch Act funds was established in 1955 and is based primarily on farm and rural population (Knoblauch, et al.). Forestry research funds are allocated under a formula established by the McIntire-Stennis Act of 1962 (National Research Council, 1990, p. 18). Each state is required to match its Hatch and McIntire-Stennis funds with one dollar of state-appropriated funding. Each state's funds are then allocated by the experiment station director.

Competitive grants are allocated by a panel of scientific peers who select research projects based on the quality of the research proposals submitted to the review panel.

Special grants are allocated by legislative mandate in Congressional appropriations bills.[2] These grants provide funding for a specific research project at a specific institution (National Research Council, 1989, p. 34).

Significant changes have occurred during the last 20 years in the relative importance of these mechanisms in allocating federal agricultural research funds. During this period, there has been a decline in the relative importance of formula funding as an allocation mechanism and an increase in the importance of special grants and competitive grants. In 1970, nearly 98% of CSRS funds were allocated to the experiment stations through the formula funding system (Figure 4). By 1992, this share had declined to 50%, with special grants and competitive grants each accounting for approximately 25% of USDA funding to the states. Real formula funding per scientist year has declined by 3% per year since 1982, while special grant funds per scientist year have grown by 8% per year. Competitive grant funding per scientist year has increased by 13% per year during the same period (Figure 5).

The distribution of special grants is roughly equal across the four regions of the United States since 1983.[3] While the Northeast region received the smallest share of total special grants (Figure 6), the four regions have received roughly equal special grants per scientist year since 1980 (Figure 7). The distribution of competitive grants across regions is much more variable, both in terms of the share of total competitive grants received (Figure 8) and the competitive grant funds per scientist (Figure 9). Such differences probably reflect the different nature of these allocation mechanisms. Since special grants are allocated during the Congressional appropriations process, it is reasonable to expect that the appropriations process would reach a consensus that would yield a reasonably equal distribution of special grants among all regions of the country. Competitive grants, on the other hand, are allocated without reference to regional distribution. Consequently, differences in institutions' ability to obtain competitive grants are likely to result in a wider variation in the distribution of competitive grants across states and perhaps across regions.[4]

Federal-State Relations in Funding Agricultural Research

As more federal funds have been allocated through competitive and special grants, the federal government's effort at matching state appropriations with formula funds has declined. In the parlance of public finance theory, the Hatch and McIntire-Stennis programs are matching grants -- the federal government matches each dollar appropriated by the states with a dollar of federal funds (up to the limit imposed by the allocation formula). As will be discussed later, matching grants are the mechanism that public finance theory suggests should be used to finance an economic good such as agricultural research which creates significant benefit spillovers outside the state in which the research is conducted. As shown in Figure 10, the federal government's matching effort has declined from 37 federal cents per state dollar in 1971 to 17 federal cents per state dollar in 1990.[5] As discussed in the next section, this trend suggests that the federal government's compensation for the benefit spillovers created by the states is declining and, as a result, the states have less incentive to invest in agricultural research.

Policy Issues Related to Research Funding

Several policy issues can be distilled from the trends in funding reviewed above. These issues involve:

- The decline in federal formula funding relative to state support.

- The role of competitive grants and special grants in funding agricultural research.

- Broader changes in federal-state relations that could have an impact on agricultural research funding at the state level.

The decline in formula funding relative to state support. The declining level of federal formula funds relative to state appropriations can be interpreted as a signal of a growing imbalance in the burden of financing research. Agricultural research is an economic good that creates benefit spillovers (externalities in the economist's terminology) outside the state funding the research.

Agricultural research benefits can spill across state boundaries in two ways. First, technology developed from research financed by one state may be adopted by producers outside the funding state, yielding benefits to farmers outside the funding state. Second, since most of the benefits of agricultural research ultimately accrue to consumers as lower prices, consumers outside the funding state often benefit from research spillovers (Ruttan, 255-58).

Because an individual state cannot capture the full benefits created by research, states have little incentive to invest in a nationally optimal level of research. One means of establishing an optimal level of investment is through the use of a matching grant from the national government to the states (the Hatch and McIntire-Stennis programs are closed-ended matching grants). Under a matching grant program, the federal government matches each dollar appropriated by the states with a specified number of federal dollars. If a nationally optimal level of investment in the spillover good is to be achieved, the matching rate of such a grant should reflect the share of the marginal benefits of research that spill across state borders. When this condition is met, the share of the cost of research paid by the federal government will equal the share of the benefits of research that spill across state borders, and the share of the cost of research paid by an individual state will equal the share of the benefits of research retained within its borders. By equating the marginal benefit and marginal cost of research at each level of government, the matching grant provides adequate compensation for the spillovers created by the states (Boadway and Wildasin).

Empirical evidence suggests that 33 to 66% of the benefits of agricultural research spill across state borders (Evenson, Waggoner an Ruttan), indicating that agricultural research should be funded through a matching grant program with the federal government matching each state dollar with somewhere between 50 cents to $2.00 of federal funds.[6] As shown earlier (Figure 10), the federal government's matching effort through Hatch and McIntire-Stennis formula funds has declined to nearly 17 federal cents per state dollar appropriated for research.

Many observers believe that the states' inability to capture the full benefits of research has caused underinvestment in agricultural research (Ruttan; Bonnen, 1985). The decline in the federal government's effort to match state funds indicates that an increasing share of the cost of agricultural research is being funded by the states. As a result, states are receiving less compensation for the research benefit spillovers they create, and the states have less of an incentive to invest in a nationally optimal level of agricultural research, thereby perpetuating, or perhaps worsening, the underinvestment problem in the future.[7]

Although public finance theory indicates that agricultural research should be funded with an open-ended matching grant in which the federal government matches each state dollar with federal dollars, the Hatch and McIntire-Stennis funds are provided as closed-ended matching grants (i.e., the federal government matches the state appropriation only up to the level established by the allocation formula). If agricultural research is not funded with an open-ended matching grant, attention must be paid to the formula used to allocate formula funds. This question arises periodically and is as old as the Smith-Lever Act of 1914, which established a population-based formula for allocating extension funds. A different variation of this formula is used to allocate Hatch funds.[8] This formula favors those states with a large number of small farms, and other formulas based on value of production or land in farms are sometimes proposed as alternatives that would better reflect the productivity of research expenditures in each state (Hodgson; U.S. Congress, p. 200). Such changes would have a significant impact on the allocation of research resources both between and within regions (Table 1).[9] The Western and North Central regions would gain formula funds under either of these alternatives, while the Southern and Northeastern regions would lose funding.

From a public finance perspective, the argument can be made that the allocation formula should allocate a larger share of funds to those states with the highest marginal product -- or the highest dollar payoff -- for an additional dollar invested in research. The practical difficulty, of course, is in finding a measure of the marginal payoff from research in each state and in finding an allocation formula that accurately represents this marginal payoff.[10]

The role of competitive grants: Public finance theory indicates that the federal government should pay a larger share of the cost of research as the share of benefits that spill across state lines increases. At some point, the federal government's share of the cost of research is large enough that nearly the entire cost of research is paid by the federal government. In such cases, research funding could be administered through competitive grants that are funded primarily with federal funds and allocated by panels of scientific peers familiar with those areas of science that will have broad applications in agriculture. Thus, public finance theory would lead us to conclude that formula funds should be used to finance research that yields a significant share of the benefits of research to the state conducting the research, while competitive grants should be used to finance research whose benefit spillovers prohibit an individual state from financing a significant portion of the research.

The public finance approach provides some useful guidelines for thinking about the role of competitive and formula funding of research. There is, however, another dimension to the research funding problem that deserves attention. Agricultural research is unlike nearly every other field of science because research problems in agriculture are generally location specific, requiring attention to particular climatic or agronomic factors that define the unique problems of a particular production region (Ruttan, p. 250-251). As a result, resource allocation decisions in agricultural research require a significantly higher level of information about local problems and conditions (Schultz, p. 16).

Economists define the cost of making decisions -- including the cost of collecting and analyzing information and the cost of negotiating group decisions -- as transaction costs (Williamson). Because agricultural research decisions require more information about local problems and conditions than do decisions in most other fields of science, its transaction costs of soliciting and transferring information from research users to research managers are much higher than other fields of science. One must ask -- what are the transaction costs of operating a competitive grant system versus the costs of operating a formula funded system, and how are these costs distributed across users, scientists, and administrators? Furthermore, are the costs of acquiring and transferring information from users to managers under a competitive grant system a serious obstacle to the establishment of a research agenda that is in concert with the needs of research users? One advantage of a decentralized agricultural research system supported primarily by state appropriations and formula funds is that such a system minimizes the transaction costs of transferring knowledge about problems from users to managers. At the same time, such a system allows research managers to communicate with users about emerging research opportunities that are relevant to their problems.

Whether these transaction costs are a serious barrier to the operation of a centralized competitive grant program is a question worth considering and a question that social scientists should be examining. Henry Wadsworth compared the priorities of the National Extension Initiatives with those of National Agricultural Research Committee and found that nearly half of the extension priorities had no counterpart in the research priorities (Wadsworth; Office of Technology Assessment, 1990). James Bonnen (forthcoming) notes that the results of the Social Science Agricultural Agenda Project found a similar dissonance between extension and research priorities. Are these observations an indication that the problems faced by research users and the research opportunities seen by scientists and research managers are not being articulated through the system? If so, can a competitive grant program operate in a way that avoids an excessive level of transaction costs and a worsening of the information transfer problem? The answers to these questions are unclear, but these questions are relevant to future decisions on research policy and deserve investigation by social scientists.

In a similar vein, what are the costs of "grantsmanship" -- defined as the costs or writing and reviewing grants -- in a competitive grant system? These costs could also be considered part of the transaction costs of operating a competitive grant system of funding. The general science community is increasingly concerned with the costs of grantsmanship in their competitive grant programs (Chubin and Hackett, pp. 25-28). Twenty-two percent of the proposals submitted to the USDA competitive grants program were funded in 1992 (Abelson), and a similar acceptance rate is reported by other funding agencies (Chubin and Hackett, p. 26). Were the scientist years embedded in the rest of these proposals a sunk cost of operating a competitive grant system and, if so, how much of a cost does it represent? Once again, these are research questions that should be examined by social scientists.[11]

The role of special grants: A third set of issues surround the use of Congressionally earmarked special grants in financing agricultural research (Marshall; Congressional Research Service, 1992a and 1992b). Critics of special grants see such grants as politically allocated funding that is distributed without regard to scientific quality. Defenders of such grants contend that special grants provide a means of directing research resources to problems faced by research users and see the legislative process as an additional means by which users can articulate their preferences for research about relevant problems (Chubin and Hackett, pp. 153-162; Browning).

What is really interesting to ponder, however, is the question of why we have seen such growth in special grants in recent years.[12] There are at least two ways in which to answer this question. The first views this growth as a consequence of a breakdown in the research system. Research users have found that the legislative process is more responsive to their problems than is the agricultural research system. In this view, the growth in special grants is a symptom of larger problems in the management of the agricultural research system and the failure of the system to respond to the needs of research users.

A second view is that interest groups and/or administrators in experiment stations are simply making a rational calculation in bypassing traditional funding mechanisms and looking toward special grants as a means of gaining additional funding. In this view, administrators know that the odds of gaining a significant increase in formula funds are slim and that even if such an increase were forthcoming from Congress, each state would receive a small increase by the time it was divided among all states. Consequently, it is rational for each state to seek special grants that it can claim as its own rather than pursue an increase in formula funding that will yield it a smaller increase in funding for its efforts.[13] Regardless of how one views the growth in special grants, the use of special grants will remain a contentious policy issue in future years. Once again, there is a need for social scientists to examine the implications of alternative funding mechanisms.

This review of the three major funding mechanisms (formula funding, competitive grants, and special grants) reveals the need to further assess the comparative advantages of each of these mechanisms. The issue is not whether one these mechanisms is the best mechanism for funding agricultural research, but is instead a question of what mix of these mechanisms can most effectively accomplish the varied objectives of the agricultural research system.

This issue can only be addressed by considering the following questions:

- What incentives does each mechanism create for state support of research and how does each of these mechanisms affect research funding at the state level?

- What are the transaction costs of operating each funding mechanism, who bears these costs, and how do these costs affect communications between research users and managers?

- What is the capacity of each of these mechanisms to allocate resources to emerging fields of science of long-term importance?

- What is the capacity of each of these mechanisms to respond to problems of immediate importance to research users?

Changes in federal-state relations: A final policy issue that deserves consideration is the changing status of intergovernmental relations in the U.S. and the impact of these changes on agricultural research funding at the state level. Recent years have witnessed a reorganization of intergovernmental responsibilities, and the cost of providing many public services has been shifted from the national government to the states (Bahl). Federal assistance to state and local governments has declined in recent years, leaving states to bear a greater portion of the cost of many programs. At the same time, an increasing number of federal mandates are consuming an increasing share of state budget resources while the federal government competes with the states for their traditional tax revenue sources (Conlan; Hamilton and Wells, pp. 87-90; Stiles).[14] As Martha Derthick has observed:

Congress loves action -- it thrives on policy proclamation and goal-setting -- but it hates bureaucracy and taxes, which are instruments of action. Overwhelmingly, it has resolved this dilemma by turning over the bulk of administration to state governments or any organizational instrumentality it can lay its hands on whose employees are not counted on the federal payroll....The fact that Congress needs the states is no guarantee, however, that it will not abuse them....In particular, there is a danger that Congress, in striving to close the gap between its large goals and its unwillingness to provide the administrative means to achieve them, will try to conscript the states. That is, it will give them orders as if they were administrative agents of the national government while expecting state officials and electorates to bear whatever costs ensue (Derthick, p. 36).

These trends are visible when examining the share of total state budgets that are appropriated to agricultural research. The experiment stations' share of the total state budgets declined from an average of 0.18% during the 1970 to 1980 period to 0.17% during the 1981 to 1990 period, and reached 0.166 in 1990, the lowest level of the last 20 years (Figure 11).[15] If this trend continues, the implication for agricultural research funding at the state level will sooon be clear. The 1990's will be an era of fierce competition for state budget resources, during which agricultural research will be at a disadvantage relative to both mandated programs and programs with wider constituencies.

Conclusions

Science funding requires a varied but integrated and adapted mix of funding mechanisms. This mix must include, but might not be limited to, formula or institutional funding and peer-reviewed, competitive grants. The type of research, its purpose (e.g., the disciplinary, developmental, adaptive, and maintenance research that are characteristic of the experiment station mission), and the structure of spillovers should control the nature and mix of the allocation mechanisms used to fund research.

The agricultural research system faces a major debate in the next decade over the level and distribution of research funds. If this debate is to have a constructive conclusion, it must focus on the appropriate roles for the full range of research allocation mechanisms, including the implications of these mechanisms for agricultural research funding at both the national and state levels. In addition, this discussion must consider both role of agricultural research policy in the broader context of U.S. science policy and the lessons that can be learned from the research funding and management practices of the larger science community.

Table 1. Comparison of Alternative Methods of Allocating Hatch Formula Funds: Share of National Land in Farms, and Share of National Farm Production, 1987.

	National Share of: Hatch Factor	Land in Farms	Farm Prod'n	(2 - 1)	(3 - 1)
SOUTHERN REGION					
Alabama	2.063	0.948	1.403	-1.115	-0.661
Arkansas	1.917	1.488	2.440	-0.429	0.523
Florida	1.856	1.161	3.198	-0.696	1.342
Georgia	2.593	1.114	2.069	-1.479	-0.524
Kentucky	3.255	1.453	1.526	-1.802	-1.730
Louisiana	1.725	0.830	0.985	-0.895	-0.740
Mississippi	1.902	1.114	1.369	-0.788	-0.533
North Carolina	3.662	0.980	2.603	-2.683	-1.059
Oklahoma	1.993	3.270	1.996	1.278	0.003
South Carolina	1.760	0.493	0.646	-1.266	-1.114
Tennessee	2.809	1.216	1.189	-1.593	-1.620
Texas	4.096	13.531	7.754	9.435	3.657
Virginia	2.393	0.900	1.168	-1.494	-1.225
REGION	32.026	28.499	28.345	-3.527	-3.681
NORTH CENTRAL REGION					
Illinois	3.784	2.958	4.687	-0.827	0.903
Indiana	3.568	1.677	2.990	-1.891	-0.578
Iowa	3.861	3.280	6.561	-0.581	2.700
Kansas	2.153	4.835	4.761	2.681	2.607
Michigan	3.375	1.070	1.871	-2.305	-1.504
Minnesota	3.449	2.755	4.172	-0.693	0.724
Missouri	3.361	3.029	2.679	-0.333	-0.682
Nebraska	2.061	4.697	4.901	2.637	2.840
North Dakota	1.414	4.182	1.608	2.768	0.194
Ohio	4.103	1.555	2.524	-2.548	-1.579
South Dakota	1.496	4.578	1.999	3.083	0.503
Wisconsin	3.438	1.722	3.609	-1.717	0.171
REGION	36.064	36.338	42.362	0.274	6.298
WESTERN REGION					
Alaska	0.611	0.106	0.013	-0.505	-0.598
Arizona	0.880	3.762	1.197	2.882	0.317
California	2.965	3.173	10.233	0.208	7.269
Colorado	1.257	3.530	2.310	2.274	1.054
Hawaii	0.627	0.178	0.448	-0.449	-0.179
Idaho	1.245	1.445	1.668	0.200	0.423
Montana	1.134	6.242	1.137	5.108	0.003
Nevada	0.626	1.036	0.184	0.409	-0.442
New Mexico	0.875	4.771	0.779	3.896	-0.096
Oregon	1.558	1.847	1.357	0.288	-0.201
Utah	0.780	1.036	0.454	0.256	-0.326
Washington	1.737	1.671	2.146	-0.066	0.409
Wyoming	0.754	3.483	0.497	2.729	-0.257
REGION	15.051	32.280	22.425	17.229	7.374

NORTHEASTERN REGION					
Connecticut	0.973	0.041	0.263	-0.932	-0.710
Delaware	0.693	0.063	0.326	-0.630	-0.367
Maine	0.975	0.139	0.298	-0.836	-0.677
Maryland	1.330	0.248	0.727	-1.081	-0.603
Massachusetts	1.157	0.064	0.250	-1.093	-0.907
New Hampshire	0.834	0.044	0.079	-0.789	-0.755
New Jersey	1.143	0.093	0.365	-1.050	-0.778
New York	3.006	0.873	1.795	-2.134	-1.211
Pennsylvania	3.823	0.816	2.262	-3.007	-1.561
Rhode Island	0.601	0.006	0.028	-0.594	-0.573
Vermont	0.847	0.146	0.276	-0.701	-0.571
West Virginia	1.478	0.350	0.199	-1.128	-1.279
REGION	16.859	2.883	6.867	-13.976	-9.992

Sources: Column 1 – unpublished USDA data; Columns 2 and 3 – U.S. Department of Commerce, 1989.

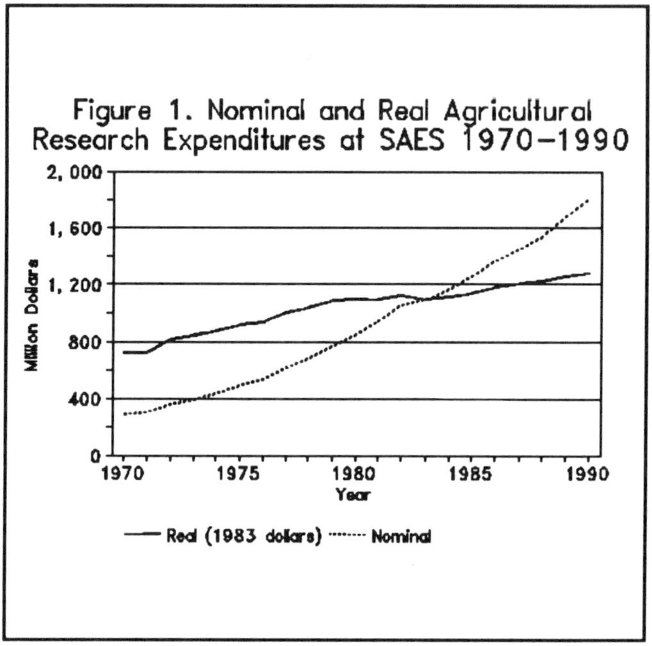

Source: U.S. Department of Agriculture; Research Associates.

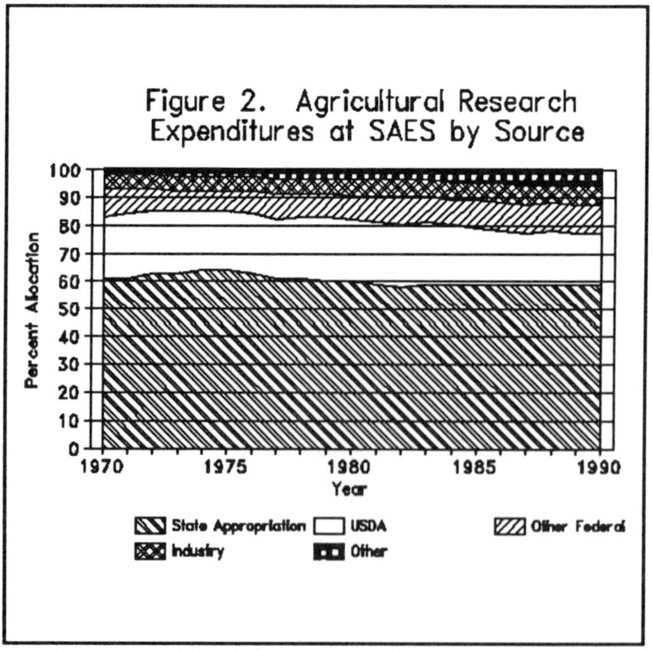

Source: U.S. Department of Agriculture.

Figure 3. Funding Per Scientist Year at SAES by Funding Source (1983 $)

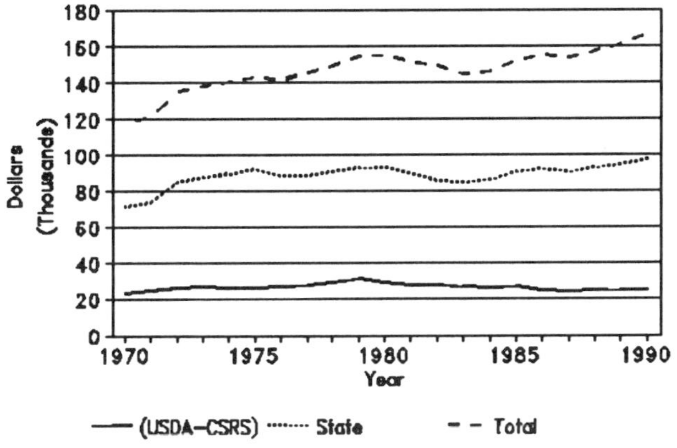

Source: U.S. Department of Agriculture; Research Associates.

Figure 4. Distribution of CSRS Funds to SAES by Allocation Mechanism

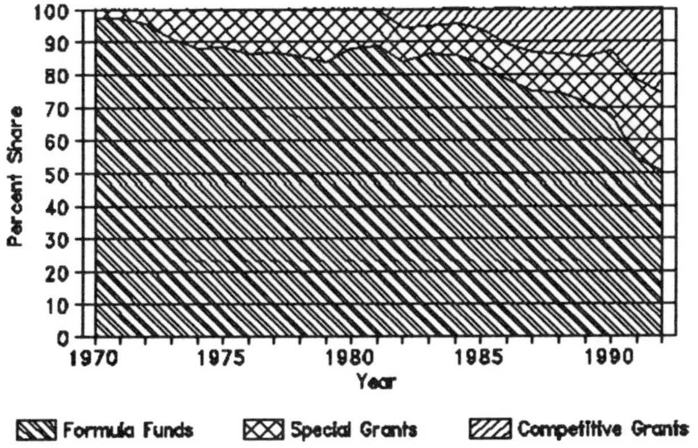

Source: U.S. Department of Agriculture.

Figure 5. CSRS Funds Per SAES Scientist Year by Allocation Mechanism (1983 $)

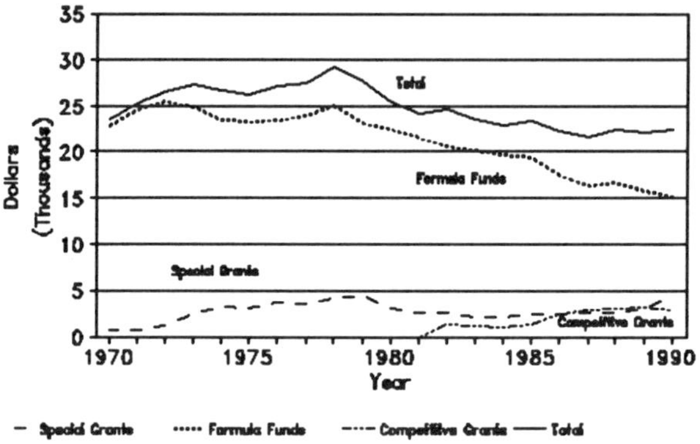

Source: U.S. Department of Agriculture: Research Associates.

Figure 6. Share of Total Special Grants to SAES Received by Region

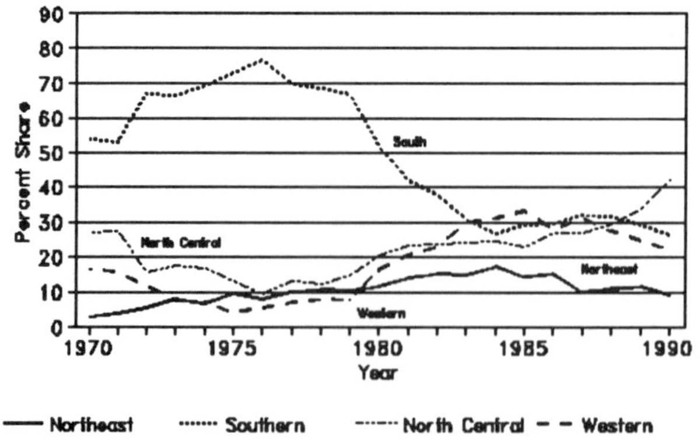

Source: U.S. Department of Agriculture.

Figure 7. Special Grants Per SAES Scientist Year by Region (1983 $)

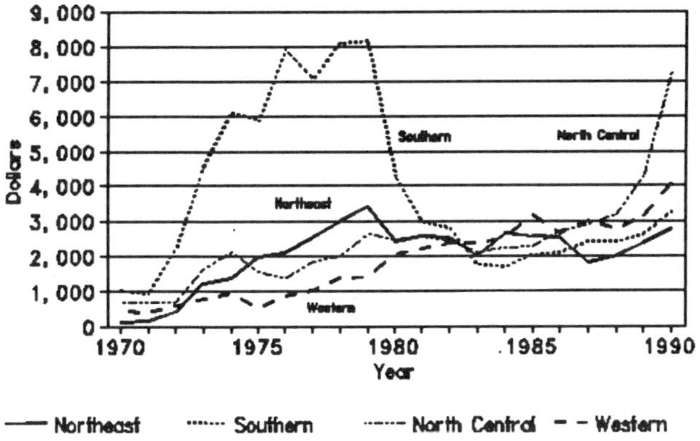

Source: U.S. Department of Agriculture; Research Associates.

Figure 8. Share of Total Competitive Grants to SAES Received by Region

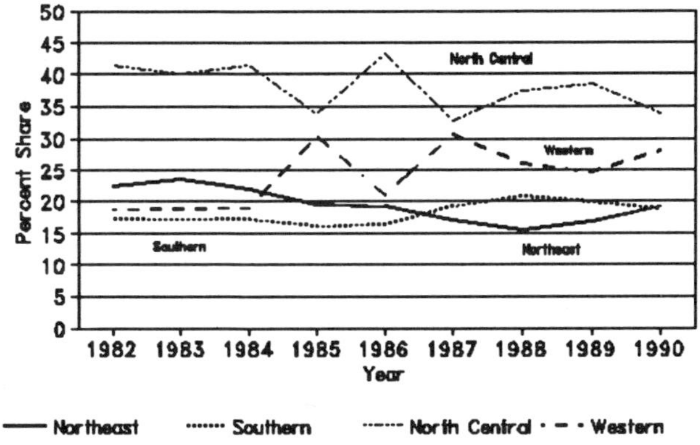

Source: U.S. Department of Agriculture.

Figure 9. Competitive Grants Per SAES Scientist Year by Region (1983 $)

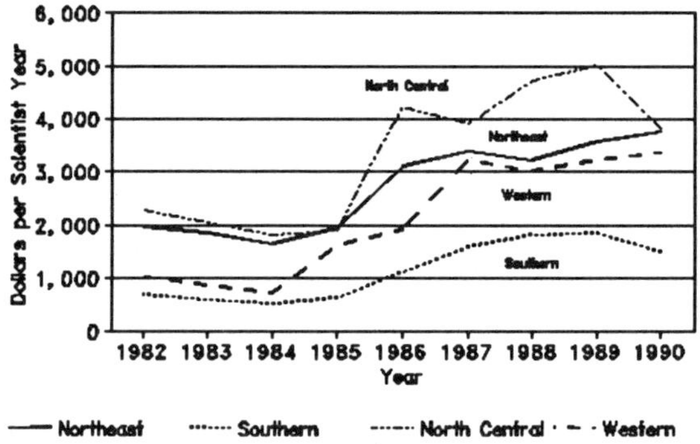

Source: U.S. Department of Agriculture; Research Associates.

Figure 10. Federal Matching Dollars to SAES Per State Dollar Appropriated

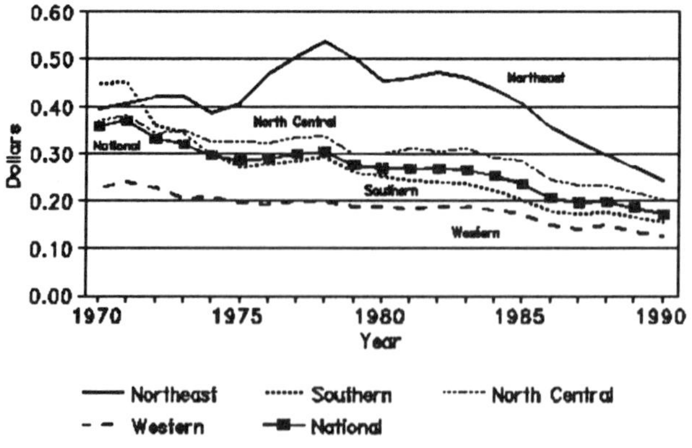

Source: U.S. Department of Agriculture.

Figure 11. Percent of Total State Budgets Appropriated to SAES Research

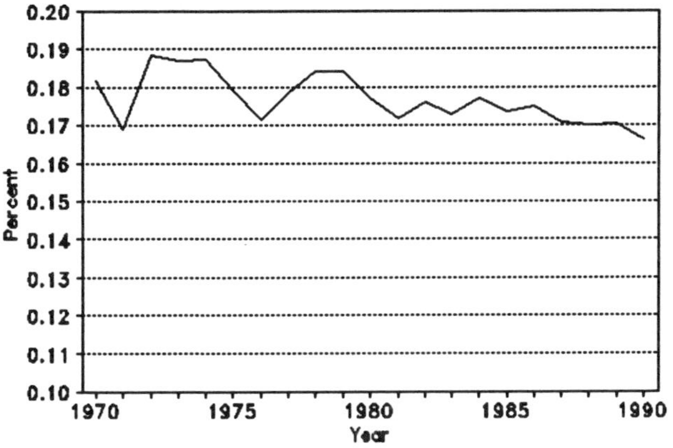

Source: U.S. Department of Agriculture; U.S. Department of Commerce.

Endnotes

1. All data used in the report refer to research expenditures at the state agricultural experiment stations. Other public and private research expenditures are not included.

2. Special grants are also known as "earmarked grants."

3. The regions used in this paper are the same as the regions used by the USDA-CSRS reporting system.

4. Available evidence suggests that wide variation exists in the ability of different institutions to obtain competitive grants. In 1988, for example, ten states obtained 73% of the total competitive grants for biotechnology research that were obtained by all experiment stations in the U.S. (Office of Technology Assessment, 1990, p. 46).

5. This ratio is calculated as total formula funds allocated to the state agricultural experiment stations (Hatch funds plus McIntire-Stennis funds) divided by state appropriations provided to the experiment stations.

6. These matching rates are calculated as MR = s/(1 - s), where MR is the optimal matching rate for financing research and s is the share of research benefits that spill across state lines (Harford; Schweikhardt). If 33% of the benefits of research spill across state lines, the federal government should establish a matching rate of 50 federal cents per state dollar (MR = .33/[1 - .33] = 0.50) to compensate states for the research spillovers created.

7. A similar problem arises when considering research benefit spillovers in an international context, since individual countries will not have an incentive to invest in an internationally optimal level of research when benefit spillovers are present. Institutional innovations are needed if workable cost-sharing arrangements for financing research are to be established at the international level (Schweikhardt and Bonnen).

8. Hatch funds are allocated to the states on the following basis: 20% allocated equally among the states; 26% allocated by rural population; 26% allocated by farm population; 25% allocated for regional projects; 3% allocated for administration (Knoblauch, et al.).

9. Hatch funding totaled $262 million dollars in Fiscal Year 1990. Thus, a change of 1 percentage point in a state's Hatch allocation factor would change its Hatch funding by $262,000. For example, Table 1 indicates that Alabama would lose approximately $292,000 (1.115*$262,000) if the Hatch formula is changed from the existing formula to a formula based on land in farms.

10. An additional issue is whether any of these formulas reflect the distribution of benefits that arises in many emerging lines of research. For example, would any of the formulas discussed above reflect the distribution of benefits that arise from research on food safety or environmental quality? The distribution of benefits from these types of research is not well understood. Social scientists could contribute to this policy debate by examining the distribution of such benefits and by developing policy guidelines for funding these emerging fields of research.

11. Existing estimates of the cost of grant writing and review indicate that such costs could be significant. Some estimates indicate that as much as 50% of research resources are spent on writing, reviewing, and reporting on grants (Office of Technology Assessment, 1990, p. 22). Other estimates indicate that one dollar of resources is spent to obtain three dollars in grant funding (Richards and Davis). Further research is needed to accurately assess these costs.

12. The use of earmarked grants has also increased in areas of science other than agriculture. A recent study of earmarking indicates that total federal funds earmarked for research in various federal departments has increased from 10 million dollars in 1980 to 708 million dollars in 1992 (Congressional Research Service, 1992a, p. 25).

13. Similar explanations of "rent-seeking" behavior have been used to analyze a wide variety of political decisions (Mueller, pp. 242-244), but have not been used to analyze earmarking of research funds. It should be noted, however, that simple rent-seeking theories cannot provide a complete explanation of the growth in special grants. It has always been true an individual state could gain an advantage by seeking special grants outside the formula funding system, yet growth in these grants did not accelerate until recent years. Consequently, other factors must have changed to permit this growth to occur. Penner and Hardin both suggest that the decentralization of power in Congress, combined with a weakening of the Congressional leadership's control of the budget process, have permitted greater access for interest groups seeking increased spending. This greater access increases the probability that such rent-seeking behavior, including the pursuit of special grants for research, will succeed.

14. Medicaid and corrections are the two fastest growing items in the budgets of many states. Much of the increase in these categories is a result of legislative or judicial mandates beyond the control of state legislatures (Gold, p. 111).

15. Since total state budgets were $571 billion dollars during 1990, a decrease of 0.01% in the experiment stations' share of state budgets translates into a loss of $57 million in state appropriations for the experiment stations.

References

Abelson, Philip H. 1992. "Agricultural Research." *Science*. No. 257, p. 1187.

Bahl, Roy W., Jr. 1990. "Changing Federalism: Trends and Interstate Variation." In *The Changing Face of Fiscal Federalism*, edited by Thomas R Swartz and John E. Peck. Armonk, NY: M. E. Sharpe.

Boadway, Robin W. and David E. Wildasin. 1984. *Public Sector Economics*. 2nd rev. ed. Boston: Little, Brown, and Company.

Bonnen, James T. 1986. "A Century of Science in Agriculture: Lessons for Science Policy." *American Journal of Agricultural Economics*. Vol. 68, pp. 1065-1080.

Bonnen, James T. Forthcoming. "Changing Roles Agricultural Economists and Their Institutions. *American Journal of Agricultural Economics*.

Browning, Graeme. 1992. "Colleges at the Trough." *The National Journal*. pp. 565-569.

Chubin, Daryl E. and Edward J. Hackett. 1990. *Peerless Science: Peer Review and U.S. Science Policy*. Albany: State University of New York Press.

Congressional Research Service. 1992a. *Trends in the Apparent Distribution of Apparent Academic Earmarks in the Federal Government's FY 1980-92 Appropriations Bills*. A report prepared for the Committee on Science, Space, and Technology of the U.S. House of Representatives.

Congressional Research Service. 1992b. *The Distribution of Apparent Academic Earmarks in the Federal Government's FY Appropriations Bills*. A report prepared for the Committee on Science, Space, and Technology of the U.S. House of Representatives.

Conlan, Timothy J. 1989. "The Disappearing 10th Amendment: Conflicting Trends, Competing Futures." *The Journal of State Government*. Vol. 62, pp. 50-55.

Derthick, Martha. 1986. "Preserving Federalism: Congress, the States, and the Supreme Court." *The Brookings Review*. Vol. 4, pp. 32-37.

Evenson, Robert E.; Paul E. Waggoner; and Vernon W. Ruttan. 1979. "Economic Benefits from Research: An Example From Agriculture." *Science*. 205, pp. 1101-1107.

Gold, Steven D. 1990. "State Finances in the New Era of Fiscal Federalism." In *The Changing Face of Fiscal Federalism*, edited by Thomas R Swartz and John E. Peck. Armonk, NY: M. E. Sharpe.

Hamilton, Christopher and Donald T. Wells. 1990. *Federalism, Power, and Political Economy: A New Theory of Federalism's Impact on American Life*. Englewood Cliffs, NJ: Prentice Hall.

Hardin, Clifford M. "Congress is the Problem." 1986. *Choices*. Premiere Issue, 1986, pp. 6-10.

Harford, Jon. 1977. "Optimizing Intergovernmental Grants With Three Levels of Government." *Public Finance Quarterly*. 5: 99-115.

Hodgson, Harlow J. 1982. *Agricultural Production and Agricultural Research Allocation Patterns*. Madison: Wisconsin Agricultural Experiment Station, 1982.

Knoblauch, H. C.; E. M. Law; and W. P. Meyer. 1962. *State Agricultural Experiment Stations: A History of Research Policy and Procedures*. U.S. Department of Agriculture, Miscellaneous Publication Number 904.

Marshall, Eliot. 1992. "George Brown Cuts Into Academic Pork." *Science*. 258: 22.

Mueller, Dennis C. 1989. *Public Choice II*. Cambridge: Cambridge University Press.

National Research Council. 1989. *Investing in Research: A Proposal to Strengthen the Agricultural, Food, and Environmental System*. Washington, D.C.: National Academy Press.

National Research Council. 1990. *Forestry Research: A Mandate For Change*. Washington, D.C.: National Academy Press.

Office of Technology Assessment. 1981. *An Assessment of the United States Food and Agricultural Research System*. U.S. Congress, Office of Technology Assessment Publication OTA-F-155.

Office of Technology Assessment. 1990. *Agricultural Research and Technology Transfer Policies for the 1990's*. A Special Report of OTA's Assessment on Emerging Agricultural Technology: Issues for the 1990's. U.S. Congress, Office of Technology Assessment Publication OTA-F-448.

Penner, Rudolph G. 1992. "The Budget Deficit: Political Monster Born of Political Pets." *KPMG Peat Marwick Legislative and Policy Developments in Washington*.

Research Associates. 1990. *Higher Education Price Indexes*. Tenth Edition. Washington, D.C.: Research Associates of Washington.

Richards, G. E. and Martha David. 1990. "Scientist Involvement in Research Funding at the Arkansas Agricultural Experiment Station." *Journal of Production Agriculture*. 3: 602-605.

Ruttan, Vernon W. 1982. *Agricultural Research Policy*. Minneapolis: University of Minnesota Press.

Schultz, Theodore W. 1985. "Agricultural Research: Canada and Beyond." In *Economics of Agricultural Research in Canada*, edited by K. K Klein and W. H. Furtan. Calgary: University of Calgary Press.

Schweikhardt, David B. 1989. *Financing Agricultural Research in a Federal System of Government: Optimal Cost-Sharing for State and National Investments*. Unpublished Ph.D. dissertation, Department of Agricultural Economics, Michigan State University.

Schweikhardt, David B. and James T. Bonnen. 1991. "Financing Agricultural Research in the Presence of International Benefit Spillovers: The Need for Institutional Coordination and Innovation." Invited paper presented at the XXI International Conference of the International Association of Agricultural Economists, Tokyo, Japan.

Stiles, William A., Jr. 1988. "Current and Emerging National Issues Relevant to Agricultural Research Funding." In *Emerging Priorities for Agricultural Research in the 1990's*. Proceedings and Minutes of the Thirty-seventh annual meeting of the Agricultural Research Institute, Washington, D.C.

U.S. Congress. 1986. *Science Policy Study -- Hearings Volume 17: Research Project Selection*. Hearings before the Task Force on Science Policy of the Committee on Science and Technology of the House of Representatives. 99th Congress, 2nd Session, 1986.

U.S. Department of Agriculture. *Inventory of Agricultural Research*. Annual issues.

U.S. Department of Commerce. 1989. *1987 Census of Agriculture: United States Summary and State Data*. Volume 1, Part 51. Bureau of the Census.

U.S. Department of Commerce. *State Government Finances*. Series GF-89-3. Bureau of the Census. Annual issues.

Wadsworth, H. A. 1989. "Opportunities for Public Policy Education in the Extension Initiatives." In *Increasing Understanding of Public Problems and Policies--1989*. Chicago: Farm Foundation.

Whims, John F.; Paul C. McVeigh; and Larry J. Connor. January 1992. *A Comparative Trend Analysis of Funding Sources for Michigan State University's Agricultural Experiment Station and Cooperative Extension Service vs. The AES and CES Units of All Other States During the Decades of the Seventies and Eighties*. Michigan Agricultural Experiment Station Special Report on the Status and Potential of Michigan Agriculture, Phase I, Report Number 4.

Williamson, Oliver E. 1985. *The Economic Institutions of Capitalism: Firms, Markets, and Relational Contracting*. New York: The Free Press.

Chapter 8

The Trends and Market for Agricultural R&D, Available Scientists, and New Scientists

Wallace E. Huffman
Iowa State University

Research and development, leading to advances in science and technology, have been and continue to be important public and private sector activities in the United States. The relative contribution of each sector has changed over time. In many developing countries, R&D continues to be primarily a public sector activity. Furthermore, agricultural technology transfers across geographic regions (or borrowing) whether across countries, regions, or states generally require at least adaptive research to make them useful to local agriculture.

In the United States, the public sector R&D activity is heavily focused on advances in science or knowledge. The new information is published in professional journal articles, scientific bulletins, and books and recorded in other materials, e.g., patents, plant variety protection certificates (PVPCs). Some of this material is protected by copyrights, patents, and PVPCs, but much is not. Historically the public sector has made innovations available to users at minimal cost. Intellectual creativity and imagination are required for university scientists to have long-term productivity. These characteristics follow partially from innate ability and partially from skills created by investing in human capital. Economic incentives and institutional factors appear to play a major role in affecting the productivity of public R&D.

Private sector R&D tends to be focused on the development of new products and processes that have the potential to be sold or licensed for profit, e.g., new genetic materials, new crop varieties, agricultural chemicals, pharmaceuticals, machinery, computer software.

Property rights in these materials include patents, PVPCs, copyrights, trademarks, and trade secrets. Establishing and enforcing property rights in intellectual material is costly. When transactions costs are extremely high, few new technologies are privately developed. In the U.S., the institution of patenting has existed for over two centuries, and during the last 25 years, new legislation, judicial decisions, and technologies of enforcement have strengthened property rights in intellectual material. This has improved the profit potential for private R&D and caused relatively rapid growth in the demand for private sector scientists and technicians. In general, skills for successful private sector scientists include technical competency with research methods (for the field of work) and an ability to cooperate across fields of science and technology and to meet deadlines.

The objective of this paper is to examine the trends and market for agricultural R&D, available scientists and new scientists. The order of the presentation of material is: (1) background information on institutions and market for scientists, (2) trends and market conditions, (3) new policy issues facing academic research, and (4) conclusions.

Background Information

Important innovations in the organization of R&D and science policy are first summarized and then historical evidence on the market for agricultural scientists is summarized.

R&D

Innovations in science policy and in the organization of R&D have affected the demand for and supply of scientists, the rate of advance of science and technology, and social welfare.

In science and technology policy, the single most important innovation took place during the 17th century. At that time the Royal Society of London and the Editor of Society's journal, the *Philosophical*

Transactions, established what has become the standard policy for intellectual property rights in scientific discoveries. The Editor guaranteed rapid publication to contributors of the *Philosophical Transactions of the Royal Society of London* and official support of the Society if an author's priority to ideas and findings were challenged. This institutional innovation established the policy that the first scientist to publish a view, discovery, or finding gets credit for the discovery. This policy established incentives for rapid communication and exchange of ideas among scientists so that a "building block approach" to advances in science could be undertaken (Committee on the Conduct of Science 1989). This policy has generally been applied in U.S. patenting. A valid patent application must show novelty, usefulness, and nonobviousness. It must also remove from secrecy the essential features of the invention or disclose the discovery in exchange for a monopoly on use for 17 years. Patent documents reference prior patents and scientific publications much the same as scientific papers cite earlier work.

Other recent important institutional innovations leading to R&D for agriculture are summarized in Table 1. Living material was initially excluded from U.S. patent protection, but in 1930, the Plant Patent Act was passed. It provided that asexually reproduced plants could be patented, provided that they met regular patent requirements. Sexually reproduced plants were not covered by patent-like protection until the Plant Variety Protection (PVPA) Act of 1970. Plant Variety Protection Certificates (PVPC) can be issued to an applicant showing that a variety is distinct, uniform, and stable. Farmers, however, are covered by an important exemption in the PVPA. Farmers whose primary occupation is growing crops for nonseed purposes can use their harvested seed for planting and selling to other, so called bin-run seed.

During the 1980s, patent protection was extended to advances in science or technology contained in living organisms. In the *Diamond v. Chakrabarty* court decision, the U.S. Supreme Court ruled that micro-organisms were patentable. In 1985, the *ex parte Hibberd* court case decision extended the Chakrabarty ruling to plants, seeds, tissue cultures, hybrid plants, and hybrid seeds. Utility patents can now be granted for innovations to both asexually and sexually reproduced crops. The property rights in plant materials differ in important ways, e.g., the nature of the innovation is not disclosed in PVPCs, and hence novelty (newness) is not an important characteristic. To be patentable, however, an innovation must involve an "inventive step" and the patent only covers the nature of the invention that is disclosed in the patent application. The first animal patent was granted in 1987 for the "Harvard Mouse," but further reviews are being conducted before additional animal patents are granted. Genetically engineered animals are, however, serving as important models for research on diseases occurring in people (*Editor*, 1992).

Science and technology policy is important, and it should promote R&D in the public and private sectors in such a way that linkages occur between advances in general science and practical problems and possibilities in agriculture (and the rest of the economy). One useful perspective of the R&D enterprise is as a multilayered set of activities that progress by applying a building block approach where needed blocks are drawn from various layers. General or mother sciences (e.g., general and biochemistry, general and molecular biology, physics, mathematics, economics) are located at the core, pretechnology sciences (e.g., plant and animal physiology, zoology and genetics, environmental sciences, entomology, agricultural economics) are located in the next layer, followed by applied sciences (e.g., plant and animal breeding, plant and animal pathology, agronomy, horticulture, animal science, farm management) and technology development and commercialization (see Figure 1).

This multilayered organization of R&D produces at its central core advances in basic or mother science that are primarily of a public-good nature and must be undertaken largely by the public sector (i.e., the research agencies of the federal government, the state agricultural experiment stations, universities). On the outer boundary, the new technology is of a type that can be largely developed and sold by private sector firms that are motivated largely by profit. In the intermediate layers, a mixture of public and private activities occur.

In a successful R&D system, activities are occurring at all levels simultaneously, advances at one layer are frequently held up because of a missing building block being "constructed" in another layer, and the stock of knowledge and technology is most likely growing because of the public-good nature of the discoveries and advances. In the short run, some independence of activity in a layer or field of science or technology can occur, but over the long term, advances in science and technology can only occur when linkages among at least some of the layers exist and function well. The primary reasons are (1) the practical problems in technology development are on the outer periphery of this R&D organization and they are frequently changing and (2) the basic advances in knowledge that are required to solve these problems are occurring largely in the core or general sciences. Thus, if practical problems are going to be solved, information or knowledge must be transferred and, most likely, adapted.

The pretechnology sciences serve as a catalyst and critical linkage inward to general sciences and outward to applied sciences and technology development (see Huffman and Evenson 1993, Ch. 2; Huffman and Just 1992; Huffman and Evenson 1992). Thus, new technology development by the private sector depends critically on advances in science undertaken in the public sector, i.e., public and private sector R&D activities should be largely complementary and not competitive activities.

Much of agricultural technology has geoclimatic specificity. A considerable amount of experimental and on-farm evidence shows that the performance of plants and to a lesser extent animals is affected by local geoclimatic conditions (Evenson 1989; Huffman and Evenson 1993; Griliches 1960). Geoclimatic conditions affect biological processes, nutrient and water availability, and type and extent of pathogenic and general environmental problems. See Figure 2 for the set of geoclimatic regions that we have found useful in our work.

This means that (1) new technology that is successful in one state or region will not necessarily outperform other technologies or spillover into other regions or states, (2) a national index of public agricultural research activity obtained by simply adding together contributions from the individual states is a very imperfect measure of useful public research affecting the productivity of agriculture in any state or region, (3) each state must conduct some of its own agricultural research (which might be only adaptive in nature) in order for its farmers to obtain new technologies that advance their competitiveness relative to farmers growing similar commodities in other states and regions, (4) potential size of the market for new technologies differs depending on their adaptability to particular geoclimatic conditions, and (5) it is generally most profitable for private sector firms to develop and sell technologies that can be intensively used in large areas and where new purchases are made regularly (say annually) (see Griliches 1960; Huffman 1992).

The biology of plants and animals is affected to different extents by geoclimatic conditions. This means that the potential for advances in science in these areas, the economics of commercialization, and the economics of the adoption by farms and effects on farm productivity, size, and specialization differ for crop and livestock technologies (see Huffman and Evenson 1992, 1993).

Market for Scientists

The R&D industry is an employer of high to very highly skilled labor. The National Academy of Science - National Research Council (NRC) has had a long history of tracking the number of Ph.D. degrees awarded by subfield (see Harmon 1978). Since 1973, the NRC has also been surveying individuals who have obtained Ph.D. degrees in science and engineering to obtain information on their field of employment, sector of employment (academic, government, industry, other), and nature of work. Huffman and Evenson (1993, Ch. 3) document the long-term trends in the number of U.S. doctorates awarded in agriculture and related fields and of scientists employed in research in the U.S. state agricultural experiment station system and USDA.

During the mid-1980s, a NRC committee examined the trends in the market for agricultural scientists and the supply of new agricultural scientists (see NRC 1988). This report showed that the total number of applied agricultural scientists increased from 14,800 in 1975 to 20,600 in 1985 (an increase of 33 percent; NRC 1988, p. 24). In 1975, 53 percent of the Ph.D. degree holders were employed in academia, 28 percent in industry (including consulting), and 19 percent in government. In 1985, a smaller share was employed in academia (48 percent) and a larger share in industry (34 percent). The share employed by the government sector remained basically unchanged over this time period.

The report also showed that there is a relatively large number of Ph.D. degree holders in agriculture-related basic sciences and employed in the United States. The number was 34,000 in 1975 and increased to 45,900 in 1985. Furthermore, the study showed in 1985 that the number of doctoral scientists employed in agriculture by field of employment exceeded by about 25 percent the number of doctorates trained in applied agricultural sciences and employed in agriculture fields (20,600/16,700). This occurred largely because of employment in agricultural R&D of individuals receiving Ph.D. degrees in the general and pretechnology sciences.

The NRC work did not involve an econometric study of the market for agricultural scientists, but it did present some conclusions about the market for new and existing agricultural scientists. First, no U.S. shortage of Ph.D. scientists in any area of basic and applied agricultural sciences was uncovered or anticipated for the decade extending beyond 1988. Second, an increasing share of U.S. employed Ph.D. holders working in applied agriculture sciences have been employed in industry, and this share was forecasted to increase. Third, an increasing share of Ph.D. level personnel who work in applied agriculture fields are engaged in non-research and teaching activities. These include marketing, management, and regulatory activities.

The study by Huffman and Orazem (1985) of the market for new Ph.D.s in agricultural economics continues to be the only econometric study undertaken of any part of the market for agricultural scientists. This is an area where considerable potential exists for future research. We need to know more about the functioning of the market for agricultural scientists.

The Market for R&D and for Scientists

Changes in the demand for agricultural scientists are caused by the salary (of a "standardized" scientist) and the revenue or real resources allocated to or available for research (and development). The production or training of new scientists is part of the R&D output. Thus, supply of new scientists is very much affected by the resources allocated to R&D.

R&D Funds

Since 1950, total U.S. private agricultural R&D expenditures have exceeded total public (USDA and SAES) expenditures on agricultural research (see Figure 3). The average rate of growth of total public and private R&D expenditures (constant dollars) during 1950-1990 was 3.9 percent. Between 1949 and 1952 there was a permanent realignment of the relative size of the USDA research activity in U.S. public agricultural research. From about 1923 to 1949, the USDA accounted for about two-thirds of the total. After the realignment during 1949-52, the USDA has accounted for only one-third of the total public agricultural research activity. The SAES system has been the dominant public agricultural research institution during the post-WWII period. For the decade of the 1980s, the average rate of growth of total agricultural R&D was lower than for the 40 year period, only 2.1 percent. During the 1980s, the private agricultural R&D expenditures grew much more rapidly than public research expenditure, 3.1 percent versus 0.4 percent. The USDA's expenditure on its own agricultural research actually decreased by approximately 10 percent during the 1980s.

These long term and more recent trends in resources for U.S. R&D are major factors in explaining trends in the U.S. market for scientists employed in agricultural fields. The demand is growing more rapidly in the private than in the public sector. The strengthening of property rights in intellectual materials during the 1970s and 1980s seems most likely to have contributed to the relatively rapid growth in private R&D and demand for agricultural scientists during the 1980s. These trends seem most likely to extend through the 1990s.

Within public agricultural research, some major shifts have occurred during the 1980s. These are summarized in Table 2. Real expenditures on crop research increased by 8.5 percent, expenditures on livestock research decreased by -2.7 percent, other public agricultural research expenditures increased by 5.4 percent. These differences in rates of growth of funds imply differences in the rates of growth of demand for public sector agricultural scientists--shifting toward crop and other research and away from livestock research. With crop research, large increases (> 18 percent) in real expenditures occurred for corn, wheat, other oil crops, and vegetables. Beef cattle research showed a large decline in public expenditures, -15.5 percent. These different rates of growth of public expenditures by commodity or area undoubtedly reflect some of the differences in the potential of new biotechnology research and extent of private sector R&D activity (see Office of Technology Assessment 1992).

For plant breeding, genetics and related research, we see very different relative involvement of the public and private sector across major crops (see Table 3). In 1989, the private sector is investing relatively heavily in hybrid corn and soybean research, but the USDA-SAES system is investing relatively heavily in wheat, small grains, and forage varietal development. The private sector is not heavily engaged in varietal development where the market is small or property rights in new varieties are relatively weak (see Huffman 1992).

Market for Agricultural Scientists

The recent trends in the market for agricultural scientists are of interest here. First, some updated information on the employment of scientists in agricultural and biological sciences is presented. These are from new tabulations prepared by the NRC (see Table 4).[1] The most recent NRC data are for 1989, and they show that the number of scientists working in agricultural fields was 7.4 percent larger in 1989 than in 1985. Furthermore, the distribution of scientists among sectors of employment has continued to shift away from academic employment and toward industrial employment. The definitions used most recently in NRC field groups are slightly different than the reference groups used in NRC (1988). The share of doctorates employed in industry increased by 2.9 percentage points between 1985 and 1989, and the shares employed in academia and government declined over the same period.

The number of scientists employed in biological science fields has grown more rapidly since 1985 than for the agricultural science field (12.9 percent versus 7.4 percent). Although a larger share of scientists in the biological science field is employed in academia than in agricultural sciences, the share employed in academia declined slightly between 1985 and 1989. In 1989, the share employed in industry was 29 percent, which is very similar to the share employed in this sector for agricultural science fields.

The supply of new Ph.D. trained individuals in applied agricultural and biological science fields fill most of the available positions in applied agricultural science fields. Thus, Table 5 contains information on the number of Ph.D. degrees awarded by these two major fields and for subfields of applied agriculture and agriculturally related pretechnology and general science for 1980-1991. The table shows a larger rate of growth in the supply of new scientists in the agriculturally related pretechnology and general science fields than in applied agricultural science fields, 14 percent versus 11 percent per year.

Turning to major subfield groups in applied agricultural sciences and comparing the recent years to those of the mid-1980s, the number of new doctorates is up significantly for the natural resource and environment group, up slightly for the animal science group, and roughly unchanged for the plants and soils group and the other fields group.[2] Turning to general and pretechnology sciences subfields, some of the trends are stronger. The number of new doctorates in molecular biology and cell biology are currently much larger than in the early 1980s and in biochemistry and human and animal genetics are currently significantly larger than during the mid-1980s. These are subfields that provide most of the graduate training for the new biotechnology research (Office of Technology Assessment 1992; Moffat 1992a, b; Gibbons 1992; Roberts 1992).

Policy Issues Facing Academic Institutions

Several new policy issues are facing universities that conduct agricultural research. First, regular CSRS-administered federal funds accounted for only 15 percent of SAES funding in 1990. Funds for research obtained from all types of contract, grant, and cooperative agreements accounted for 25.3 percent of SAES funds in 1990. A new USDA-CSRS competitive grants research program was initiated in 1977, and since 1985 when biotechnology research was included in the type research covered by competitive grants; competitive grant funds have grown. In 1990, $42.5 million were in the Competitive Grants Program and it has about doubled in size by fiscal 1993. The share of CSRS-administered funds that is allocated to competitive grants seems likely to increase. Also, competition for other research money seems likely to increase during the 1990s.

The greater emphasis on competition for research funds in agriculture appears likely to raise the quality of science required to compete successfully for these funds. This will strengthen upstream-science ties. Also, new Ph.D. recipients from upstream fields seem likely to continue to be an important source of new scientists for applied agricultural science fields in universities. Furthermore, biotechnology programs draw heavily from upstream sciences, e.g., molecular and cell biology. Thus, U.S. academic employment prospects for "traditionally trained" agricultural scientists appear bleak for the decade of the 1990s.

Second, the exemption that permits universities to require retirement at age 70 is about to expire. If other policies are unchanged, this will increase the average age of faculty at retirement and slow the opening of positions due to retirements and further weaken the academic market for new Ph.D. scientists during the 1990s.

The universities will need to substantially change faculty review and compensation policies in order to keep faculty retirement decisions from becoming a major personnel problem. The positive aspect, however, is that people are living longer and in better health on average. Thus, for many individuals, human capital investments can payoff in the labor market over a longer time period. Universities will need to develop stronger economic incentives for faculty improvement through sabbatical and re-training leaves. Short-term leave policies could be used by administrators to obtain more research flexibility for new topics. New university policies for senior faculty will need to incorporate stronger economic incentives for retirement. Nominal faculty salaries and real faculty compensation cannot be permitted to rise forever after age 70.

Conclusion

The total number of agricultural scientists continues to grow - but slowly. An increasing share of these scientists are being employed in the private sector. An examination of the research needs for private sector employment and comparison of those required for academic positions is needed. Newly trained scientists for the private sector and cooperative research ventures between the public and private sectors seem likely to become more important in the future and to be an important part of technology transfer and commercialization.

Agricultural research in universities is facing new policies for the 90s. These include greater emphasis on competitive grants, more cooperation with the private sector, and removal of the mandatory retirement at age 70. These promise to bring change in the economic incentive facing scientists and institutions.

Endnotes

1. My field classification scheme for Table 5 does not match any general field classification scheme for the National Research Council. Furthermore, the field classification schemes for Table 4 and 5 are different.

2. During 1980-89, the share of the doctorates in the applied agricultural science field that was awarded to individuals having temporary visas was 32.0 to 33.5 percent, but during 1990 and 1991 the share has been significantly larger – 40.0 percent in 1991.

References

Committee on the Conduct of Science. 1989. *On Being a Scientist*. Washington, D.C.: National Academy of Sciences.

Editor. 1992. "Animal Models Point the Way to Human Clinical Trials." *Science* 256(May 8, 1992):772.

Evenson, R. E. 1989. "Spillover Benefits of Agricultural Research: Evidence from U.S. Experience." *Am. J. Agr. Econ.* 71:447-452.

Gibbons, A. 1992. "Biotech's Second Generation." *Science*, 256:766-768.

Griliches, Z. 1960. "Hybrid Corn and the Economics of Innovation." *Science* 135(1960):275-280.

Harmon, L. R. 1978. *A Century of Doctorates*. Washington, D.C.: National Academy Press, 1978.

Huffman, W. E. 1992. "Economics of Intellectual Property Rights in Plant Materials." Paper, Dept. Economics. Iowa State University, Ames, IA.

Huffman, W. E. and R. E. Evenson. 1992. "Size, Specialization, and Agricultural Growth: How Public Policy Enhances Scientific and Economic Progress in Agriculture." Paper, Dept. Economics, Iowa State University.

Huffman, W. E. and R. E. Evenson. Forthcoming 1993. *Science for Agriculture*. Ames, IA: Iowa State University Press.

Huffman, W. E. and P. Orazem. December 1985. "An Econometric Model of the Market for New Ph.D.s in Agricultural Economics in the United States." *Am. J. Agr. Econ.* 67:1208-1214.

James, N. I. "Survey of Public Plant Breeding Programs in the United States, 1989." *Diversity*. Vol. 6, No. 2, 1990, pp. 32-33.

Kalton, R. R., P. A. Richardson and N. M. Frey. 1989. "Inputs in Private Sector Plant Breeding and Biotechnology Research Programs in the United States." *Diversity*. Vol. 5, No. 4, pp. 22-25.

Knudsen, M. and L. Hansen. November 1991. *Intellectual Property Rights and the Private Seed Industry*. USDA-ERS, Ag. Econ. Report, No. 654.

Moffat, A. S. May 1992a. "High-Tech Plants Promise a Bumper Crop of New Products." *Science*. 256:770-771.

National Research Council, 1988. *Educating the Next Generation of Agricultural Scientists*. Washington, D. C.: National Academy Press.

Office of Technology Assessment. August 1992. *A New Technological Era for American Agriculture*. OTA-F-474. Washington, D. C.: U.S. Government Printing Office.

Roberts, L. April 1992. "DNA Fingerprinting: Academy Reports," *Science*, 256:300-301.

Regulatory Impacts on Agricultural Research

Chapter 9

Environmental Performance of Agricultural Chemistry: The Role of FIFRA

Robert D. Weaver
The Pennsylvania State University

By any definition, regulation of economic activity constrains and directs that activity toward chosen goals. As an alternative, tinkering with market incentives can also redirect economic activity by redefining what is profitable. Either approach incurs indirect costs in both the short- and long-run. As has become increasingly apparent in many of the world's economies, these costs may be substantial. As profitability is redefined, the mix of the supply of goods from current technologies will shift, redefining employment of labor and other resources as well as the incomes and returns to those inputs and technologies. Of equal import, as expectations of profitability in the longer term are redefined, the incentives and opportunities for innovation are directly altered and the economy is set off onto a fundamentally different course. Going one step further, in a world of multiple, competing economies, the competitiveness of the regulated or controlled economy is altered.

These general principles are well understood by many people today, in a way which was beyond our collective experience three decades ago. The evolution of this understanding is attested to by the case of regulation of environmental performance of agricultural chemistry. As a major force affecting both private and public sector research activity, this regulation has fundamentally reshaped the level of effort and directions of basic as well as applied research, the nature and scale of innovations pursued, and the profitability of research management decisions. The objective of this paper is to present an economist's view of the corner stone of regulation directed toward environmental performance of agricultural chemistry. In the process, the paper will reassess the need for and means for redirecting research activity to achieve goals defined through social and political processes.

Incentives for Innovation in Environmental Performance in a Market Economy

To begin, most economists would argue that the role for any economic regulation must follow from a consideration of the performance of decentralized market forces. By understanding how and why those market forces fail to satisfy social and political goals, a rationale is established for a government role. Historically, the need has been recognized for some form of government action to ensure and protect incentives for innovation. In market oriented economies, patents have been adopted to establish property rights which allow inventors to exclude others from the use of innovations. Given this right, the inventor can effectively require payment in exchange for access to the benefits of the invention. Importantly, patents establish a market for the benefits of the innovation. In abstract, a patent grants an exclusive right to supply the invention to the market. In practice, all inventions face competition from a multitude of existing and future substitutes. As in any market, competing supplies and demands in the market for the invention's benefits determine the price or value of those benefits and this defines the profit generated by the invention.

In practice, similar results have been found where other forms of protecting the right of access to the benefits of an invention are used. Maintenance of trade secrets or market saturation have also proven viable strategies. Economists find patents a good example of an effective role for government that strengthens and catalyzes market processes rather than stifling them or replacing them with government bureaucratic processes. The role for government in ensuring environmental performance of agricultural chemistry must be based on an understanding of why market forces fail to generate adequate incentives to achieve social and political goals.

A basic requirement for the existence of a market for any good or characteristic of a product is that rights to consumption of the good can be defined and consumption can be controlled. It is within this realm that the environmental performance characteristics of agricultural chemistry fail to provide the basis needed for market forces to stimulate their supply.

Consider the case of field crop chemistry. Two general types of effects can be defined to result from the field application of an agricultural chemical: 1) the direct agricultural productivity effects and 2) the environmental effects possibly occurring both on and off the user's farm. As an input into a crop production process, an agricultural chemical contributes to increased output and enhances the productivity of other inputs. These direct effects generate benefits which can be *appropriated* or realized by the farmer using the chemical. In addition to such direct effects, the environmental performance of an ag chemical involves many effects that are realized by non-users or off site on non-targeted species or physical systems. These *external effects* may include soil, water, wildlife, and human health effects. This type of environmental effect can be thought of as external to the original use since it typically occurs at sites that lie beyond the scope of the direct effects of use. By definition, the user's production process is not affected by these external effects, implying the effects should be ignored by the user in any profit oriented decision to use the chemical. For this reason, the user's consideration of personal profitability of agricultural chemistry use can not be expected to result in a demand for environmental performance of a product.[1] Since in this sense no margin of profit is generated for the user by enhanced environmental performance of the agricultural chemistry, no budget is generated from which the user could pay for the enhanced environmental performance of a product. In the absence of ability, and therefore, willingness to pay for environmental performance by users, no margin of revenue is available to privately finance the supply of environmental performance.

Despite the farmer's unwillingness to pay for environmental performance, substantial social value for such characteristics of agricultural chemicals is widely recognized and implicit in the political support for FIFRA.[2] The divergence that exists between this social value and the private return to R&D focused on environmental performance of agricultural chemistry represents a key rationale for government intervention. This divergence means that market processes will fail to generate levels of environmental performance that are demanded by society. It is this failure of market forces that motivates a role for government. Intervention to stimulate R&D on environmental performance to achieve levels that reflect its social value can take many forms, one of which is regulation through standards and market gatekeeping.

Innovation in environmental performance requires expanding the scope of research and development to consider what must be interpreted as a new set of product characteristics. However, if users can not typically be expected to be able or willing to pay for environmental performance, how can that expanded scope of R&D be financed? Unfortunately, the solution of patent protection is not viable for product characteristics such as environmental performance. The protection of patents relies upon tangibility of the invention in a form that allows the use of it to be controlled. While it could be argued that environmental performance characteristics are linked directly to tangible agricultural chemical products, the correspondence is not one-to-one. The benefits of environmental performance may go well beyond one product, and typically would go well beyond the initial user. The only tangible aspect of environmental performance is, in fact, the knowledge of that performance, or equivalently the data, or research results that establish its nature. The challenge of stimulating innovation in environmental performance centers around the "production" of this type of data.

More specifically, the conclusion might be drawn that one solution to the problem of catalyzing research and development, or innovation in the environmental performance of agricultural chemistry is the establishment of intellectual property rights over the knowledge of that performance. However, for the case of environmental performance, existence of such rights can not be expected to stimulate R&D in the absence of a means of financing the R&D. For most innovations, the appeal of the patents and the property rights they establish is that they result in a basis for control of access and establish the feasibility of financing from the sale of use rights. In the case of environmental performance, the intellectual property is data or research results which represent economic goods for which access to use is notoriously difficult to control.

FIFRA as a Regulatory Strategy for Enhancing Environmental Performance

In a market economy, the demand for inputs follows from users that find their use to be profitable. The supply of inputs is likewise governed by profitability of their provision given the cost of manufacture and potential sales revenue. By design, a market mechanism effectively ensures that the right amount of such tangible inputs are exchanged at a price which approximates the unit cost of manufacture. In contrast, market economies are not adept at managing intangibles such as environmental performance. FIFRA follows the precedent of many regulatory bills by establishing standards for certain external effects of agricultural chemical use such as environmental performance.

In particular, FIFRA requires that agricultural chemicals be registered through a process that factually establishes their environmental performance through presentation of scientifically defensible data describing that performance. Beginning in 1972, FIFRA formally recognized the value of environmental performance data. The 1972 legislation prohibited use of trade secret or confidential data by post-patent imitators. Non-trade secret data could be cited if an offer of reasonable compensation were offered by the imitator. In 1975, the duty to pay for citation of data was amended to apply only to data submitted after January 1970. In 1978, the right of exclusive use of trade secret data was withdrawn and arbitration responsibility was transferred from EPA to private arbitration. In place of unlimited exclusive use, duty to pay for 15 years after data submittal was established for data submitted between January 1970 and September 1978. For data submitted after September 1978, all data was granted exclusive use for 10 years and a duty to pay for citation was established for 5 years after the exclusive use period. For data developed in response to call-ins, joint financing by registrants was required.

Operationally, FIFRA remains the federal solution to achieving environmental performance of agricultural chemistry. Interpretation of FIFRA's operational role hinges on the recognition of the change in social standards which it reflects. Prior to FIFRA, R&D in agricultural chemistry led to new chemistry which was marketable only if farmers were convinced its use was profitable. The feasibility of marketing a new product was, therefore, based almost entirely on the internal effects of the new chemistry, that is its contribution to farm productivity and profitability. In contrast, FIFRA focuses on the external effects of the use a new product. FIFRA represents a mechanism for ensuring that the external effects of agricultural chemicals meet certain standards. The implication for R&D is that the scope of effort required to market a new agricultural chemical was expanded from that which would be necessary to establish farm productivity to that necessary to establish environmental performance as well.

From an economic perspective, FIFRA plays two significant roles in the agricultural chemical market. The primary economic function of FIFRA is to establish a gatekeeping process over the right to market new agricultural chemicals or to continue marketing of chemicals that have been recalled under FIFRA 3c2B. The second role played by FIFRA follows from its regulation of environmental performance data that constitutes evidence that a product meets environmental performance standards. Through these regulations, FIFRA establishes a registration process to establish that environmental performance standards have been met.

The registration process established by FIFRA represents a market permitting process that grants rights to market products when environmental performance data supports the conclusion that they meet standards. In the absence of regulations such as FIFRA, the feasibility of bringing a new product to market would be determined by markets for its internal effects such as farm productivity and profitability, in combination with patent protection. The role of FIFRA is to add another gate to the path that leads to the market.

The second role of FIFRA represents a radical departure from the tradition of patent law. Specifically, FIFRA establishes mandatory licensing of the environmental performance data that supports claims that external effects fall within prescribed standards. Patents may be viewed as a voluntary mechanism through which the innovator may establish protection for an invention, in return for revealing information concerning the invention. In contrast, FIFRA currently mandates that results of environmental performance R&D be released by the innovating firm after ten years of exclusive use. In simple terms, this amounts to mandatory licensing of trade secret information. While for five years following the exclusive use period a duty to pay (a license fee) is established, thereafter FIFRA establishes the license fee to be zero.

Given that R&D to establish mandated environmental performance can not be financed from patent proceeds, this mechanism for compensation would seem to serve as a means for financing such

R&D. During the exclusive use period, innovators could license their data through negotiated contract. During the duty to pay period, licensing for fee could also generate revenue that could be interpreted as a return to innovation in environmental performance. However, the effects of such a financing mechanism on the level and direction of R&D effort depends ultimately on the R&D processes focused on marketable product characteristics and their relationship among those processes and R&D processes dedicated to environmental performance.

Imperatives for a Mechanism for Financing Environmental Performance

The salient characteristics of the R&D process determine the feasible means of both the management and financing of that process. If the process were project focused, linear, and subject to little uncertainty of success, then simple time and cost accounting for each project would accurately describe the effort preceding marketing of a new product. Management of each stage could be pursued independently of future stages and the economics of the process would differ little from a standard construction project. For each project, returns could be compared with costs to assess the profitability of project investment.

If, instead, the process of R&D is more general in focus, if progress is not continuously related to time and effort, and if substantial uncertainty persists throughout the process, then the conduct of R&D is more accurately perceived of as the operation of a complex engine that generates opportunity. The value of this opportunity requires expert insight and continuous investigation if it is to be recognized and appropriated by bringing its commercial implications to market. Within this type of context, the R&D organization cannot be viewed as one which simply implements a set of distinct projects each of which may be financed independently. Instead, the R&D organization must be viewed as a highly tuned engine composed of interacting teams of human and physical resources.

Most importantly, in this case, analysis of profitability must be more broadly focused. Over time, the R&D organization must be viewed as generating a stream of effort and expense which may periodically result in commercially valuable discoveries. In this sense, the revenue flow from commercialized discoveries must be viewed as returns to a more broadly defined stream of effort and expense by the R&D organization.

A further complication in financing of R&D is raised by the unattractiveness of external funds as a means of financing. Financing of product oriented R&D requires a stream of funds capable of financing the R&D organization's on-going activity. Successful commercialization requires a high level of secrecy during the discovery and development phases. This imperative limits the nature and extent of external financing available and attractive to the R&D organization. An important implication is that revenues from commercialized products are heavily relied on to provide funding for on-going R&D. In combination with exigencies to maintain stable levels of growth and return to equity, finances available for R&D are necessarily limited.

Standards for the environmental performance of agricultural chemicals leads to two important effects on R&D: 1) expansion of the scope of R&D and 2) reduction in the probability of marketable discovery and increase in the cost of marketable discovery. The environmental performance standards established by FIFRA are not simply warranty standards requiring proof of particular product performance claims. Instead, FIFRA's regulation of environmental performance requires an expansion of the scope of R&D objectives to include achievement of the standards for external effect performance. In the absence of concern for external effects, the discovery phase of R&D would involve only a search for biological activity for which a profitable market demand can be predicted. Product oriented R&D focuses on identification of potential candidate compounds and screening of the biological performance of those compounds. In contrast, where environmental performance is also of interest, the identification and screening process must be expanded in scope to include consideration of both internal effects of value to the farmer as well as external effects of potential concern. How this expanded scope of enquiry in the R&D process is operationalized varies by corporation. Importantly, environmental performance R&D activities must ultimately interact with product oriented R&D focused on biological activity valued by the farmer. At a scientific level, considerable interaction may result from joint use of basic knowledge of chemical properties and performance of biological systems.

Statements such as "On average it takes 8 years and 10 million dollars to produce a new agricultural chemical" suggest that the R&D process underlying new products is 1) product specific in focus, 2) linear and 3) highly predictable. This conceptualization also suggests that the R&D process is demand driven, starting with a specific unfilled need. In fact, this conceptualization of the R&D process is misleading. Though R&D progresses through time, the rate of effort and expenditure may be variable and discontinuous. Further, progress is highly variable, uncertain, and it is not continuously related to time, effort, or expenditure.

In the absence of comprehensive theories of biological performance, agricultural chemical R&D typically begins in a discovery phase where objectives are only given general focus to determine general forms of biological activity. Potentially active chemistry is identified, acquired, and its biological activity assessed. Feedback to chemical identification continues. During the process, numerous forms of biological activity may be identified and pursued, only some of which may be related to initial general research objectives. The biological activity of specific compounds are further explored in a development phase. Here, investigation and development of the precise form and extent of commercial viability of the compound becomes an objective. However, discovery of a basis for non-viability may have equal ultimate value for R&D or as a basis for an entirely different product.

From this perspective, the focus of R&D processes in the agricultural chemical industry is not accurately described as involving a set of product specific projects. Instead, the R&D process involves a sequence of effort and expense that is more generally directed at increasing the probability of discovery. When a narrowing of the extent of uncertainty is achieved, progress is made. The effect of environmental performance standards is to force innovators to establish not only the biological performance of an active ingredient in controlling a particular target pest on the farm but also the external biological and physical performance of the ingredient. The nature of commercially successful agricultural chemical R&D processes and the interaction between specific biological activity and more general environmental performance requires that environmental performance R&D is not simply an activity added-on to product oriented R&D, but rather is an integral part of the both discovery and development phases.

A second basis for interaction between product oriented and environmental performance oriented R&D must be noted. Specifically, interaction may be induced from two types of constraints faced by the corporation. First, R&D is limited by available human and physical scientific resources. These resources are in scarce supply at the market level and are costly to assemble. Further, the productivity of these resources takes time to evolve and depends critically on management decisions which direct and hone the package of available resources into an effective and creative team. It follows that to pursue environmental performance R&D resources must be directed from product oriented R&D. Second, financial resources limited by the exigencies for internal funding place a further significant constraint on R&D activity. Again, pursuit of environmental performance must come at the cost of diverting resources from product oriented R&D. Within the context of these constraints, R&D concerned with effects environmental performance can only be achieved by reduction of product oriented R&D activities. From this perspective, the imperative for financing environmental performance is clearly established.

Financing Environmental Performance R&D

In the absence of any regulatory standards for environmental performance of an agricultural chemical, the absence of a market for such performance characteristics imply that such external effects would be ignored by the innovator. Environmental performance R&D expenditures would earn no return and would not be attractive to privately finance. FIFRA introduces mandatory standards for certain external effects which must be met as a basis for securing the right to market. Nonetheless, the absence of a market for environmental performance characteristics or an alternative means of financing environmental performance R&D implies that environmental performance R&D can only be financed out of patent returns earned for product characteristics. Because R&D is financed internally from limited earnings and in the absence of any market value for the results of environmental performance R&D, internal financing of environmental performance R&D necessarily reduces funds available to finance both product oriented and environmental performance R&D.

FIFRA steps beyond the voluntary nature of patent law by requiring that firms share scientific results from environmental performance R&D with competitors after an exclusive use period. This constitutes mandatory licensing and FIFRA's requirement for compensation can be interpreted as requiring that licensees pay for access to such data. The question of what form data compensation should take is, therefore, a question of what is the best form of compensation for a mandatory license for use of the environmental performance data. An economically sound means of financing environmental performance R&D is needed and data compensation, interpreted as a mandatory license fee, represents an important opportunity for achieving this end.

Endnotes

1. The nature of the market for environmental performance or any other external effects of new agricultural chemicals is such that the simple economics of R&D for go awry. In contrast to protection offered to the results of R&D by patents for productivity effects, patents offer no protection to claims of the absence of specific types of environmental performance or other external effects. Farmers are, in general, not willing to pay for external performance characteristics. While some degree of altruism could be expected to lead farmers and research organizations to have some concern for external effects, the economic principles of market economies assume external effects will be ignored by decision-makers. Intuitively, by ignoring external effects the farmer limits his decisions to effects which he can observe. This limited consideration is desirable from both the farmer's and society's perspective since it eliminates errors in decisions based on imperfect information concerning external effects. The conclusion can be drawn that no market exists among users for external performance characteristics of new agricultural chemicals.

2. While farmers can be assumed to be unwilling to pay for environmental performance that does not contribute directly to their profits, evidence exists that farmers have strong interests in environmental performance and may be willing to pay for it as a result of altruistic motivations, see Weaver (1992).

References

Weaver, Robert D. 1992. Altruism vs. Hedonism as Factors Contributing to the Adoption of Conservation Practices. Staff Paper #226. Department of Agricultural Economics and Rural Sociology, The Pennsylvania State University

Chapter 10

APHIS and Agricultural Research

Dr. Rod Townsend
Pioneer Hi-Bred International, Inc.

Regulatory oversight of agricultural research is a controversial issue. Any assessment of the impact of regulations is strongly influenced by one's own perspective. This author's prospective is based on experience in managing the regulatory affairs section of a seed company (Pioneer Hi-Bred) that has a huge research program aimed at developing new varieties of the major forage, grain and oilseed crops. Our plant breeders make hundreds of thousands of crosses and selections each year. Those crosses and selections are usually made in the field, not in the greenhouse and new experimental varieties are tested on the farm, side by side with current top performing products.

We are excited by the possibilities that biotechnology presents to develop improved crops that can be grown with fewer chemical inputs and yield more nutritious foods and feeds. Pioneer has made a significant commitment to incorporate the new genetic technologies into the process of crop improvement. By new technologies, I mean those methods based on recombinant DNA manipulations that give rise to what are commonly called "genetically modified" or "transgenic organisms."

As a consequence of embracing these new technologies, we and others in the seed industry have found ourselves at the very center of the debate over environmental release of genetically modified organisms. We have become subject to regulations that are likely to have a profound effect on the future development of our industry and of the nation's food supply.

However, it was not so long ago that I was a researcher in the public sector. My formal training is in plant pathology, and I spent 15 years conducting research in plant pathology and molecular biology at a well-known public research institute in England. I still maintain working relationships with the Agricultural Research Service (ARS) and university collaborators, so I think that I may also have some insight into the impact of regulations on the academic community.

It was from that rather remote European vantage point that my colleagues and I watched, both fascinated and somewhat concerned, as Steve Lindow from UC-Berkeley, and the now deceased Advanced Genetics Sciences Company, battle with Jeremy Rifkin for the right to field test the infamous ice-minus bacterium in the early 1980's.

Those early trials had a momentous impact on regulatory policy. It is the general consensus that it was EPA's October, 1984 decision to require notification for all small scale releases of genetically engineered microbial pesticides, under the *Federal Insecticide, Fungicide, and Rodenticide Act or FIFRA*, that precipitated the charge by other federal agencies to grab a piece of the virgin regulatory territory that biotechnology represented. It soon became clear that in applying their regulations, the agencies were not necessarily going to distinguish between academic research and research for commercial application.

As an academic researcher, I, like most of my colleagues, reacted unfavorably to the prospect of regulation. We saw it as a move to restrict scientific research and a potential challenge to our "academic freedom". After all, I had a stack of papers to say that I was a recognized authority in my field and a bunch of bureaucrats was not equipped to second guess my judgments about what was safe research.

What was so special about recombinant DNA technology anyway? Wasn't it simply another step in the continuum of technologies, including hybridization, mutagenesis, and tissue culture, that have been safely employed to develop new crop varieties? Even if there were risks, scientists had proven themselves perfectly able to regulate their own activities -- compliance with the NIH guidelines had demonstrated that. Surely it was industry that needed regulations to keep it on the straight and narrow path -- I had no commercial incentive to cloud my judgment.

Furthermore, it appeared scientifically untenable that a microbe could be subject to the same jurisdictional authority as a chemical pesticide. I didn't have the financial resources of an agrochemical company. My budget was already stretched and any additional demands made by generating data for regulatory compliance could only detract from the quality of my research program.

Well, suffice to say that I found myself on the other side of the fence a couple of years later. In fact, I found myself on the other side of the Atlantic conducting molecular biology research for a multi-national American seed company. By that time, the regulatory agencies had also finished carving up biotechnology amongst themselves and had published the final version of the *Coordinated Framework for the Regulation of Biotechnology*.

In the industrial setting, I began to see regulations in a rather different context. Far from being regarded as a threat, the framework was given a qualified welcome by industry; and while certainly not viewed as perfect, it was seen as at least providing an umbrella of oversight for our research activities. The need for that umbrella had suddenly become critical as transformation techniques were developed for a number of different crop species, allowing them to be genetically engineered with genes of potential agronomic significance. With millions of dollars at stake, a stable product research and development environment is essential, and *appropriate regulations* help to provide a stable environment. The need for appropriate regulation should be emphasized since overly burdensome regulation could certainly inhibit research and development. As an example, the German "Gene law" has driven researchers and companies to establish themselves in other countries where there are more supportive regulatory climates. In the U.S., while industry has not always agreed with federal policy, they have rarely adopted an adversarial approach, instead they have sought to influence the course of policy making through a process of discussion. Much of that discussion has involved educating regulators about the industry and how the new technology has been applied.

So complete was my own conversion to the industrial perspective that I was working full-time in the area of regulation by 1989. In this paper, I would like to review from my perspective the impacts of these regulations on agricultural research in the U.S. My experience is largely with transgenic plants, so in the following review I intend to confine myself to the impact of regulations in the area of crop improvement. There are others who are better equipped to discuss issues related to microorganisms.

The Regulatory Framework: An Industry View

In considering the regulatory oversight of transgenic plants, we have the advantage of being able to analyze the application and impact of a regulatory system that has been in place for a relatively long time, and under whose jurisdiction the most field research has been conducted. The Coordinated Framework gave USDA-APHIS a substantial slice of territory; namely the regulation of importation, interstate movement and field testing of transgenic plants. APHIS regarded this as an extension of their responsibilities for plant quarantine control. Others have suggested that APHIS went much further -- stretching the *Federal Plant Pest Act* almost to breaking point with the inclusion of the new Part 340.

Part 340 places restrictions on the introduction, that is the movement within the United States or release into the environment of regulated articles, irrespective of whether the introduction is performed for academic research or commercial gain. A *regulated article* is any organism that is a plant pest or has been genetically engineered using a plant pest or part of a plant pest. A key issue in gaugin the impact of the regulations is this definition of a regulated article. The idea that a non-coding sequence from *Agrobacterium* or cauliflower mosaic virus can retain the potential to be a plant pest when cloned into a tomato plant certainly stretches the limits of scientific credibility. Nevertheless, the APHIS implementation of the coordinated framework has led to a permitting system that incorporates such a scope of definition and has been strongly supported by industry -- why?

Industry learned two very valuable lessons from the ice-minus experience. The first was the need to stay out of court. The APHIS review system includes an Environmental Assessment, that in all cases to date has led to a Finding of No Significant Impact. This review effectively ensures compliance with the requirements of the *National Environmental Policy Act* or NEPA, a favorite tool of litigious activists that figured prominently in challenges to the ice-minus experiments.

The second and most important lesson was the public concern about the technology and, in particular, the perception that something dangerous was being done in their back yard. However, experience indicates that where scientists took the time to inform people about the technology in ways they could understand it, talked openly about the trials and their expectations, and particularly how the consumer might benefit, much of the concern disappeared. Public confidence was further enhanced when they knew that an organization with more local credibility, such as a state's department of agriculture, was involved in the review of safety precautions.

Public notification and state participation in the approval process were the keys to acceptance, and APHIS, through its strong links with the states, was in the best possible position to disseminate information to the local level. The Agency's policy of publishing a notice in the *Federal Register* of each application received and of every permit issued, coupled with the provisions of Part 340.3, that require state notification and review, have been key features of APHIS' oversight.

The first field test of transgenic plants was conducted by Rohm and Hass Company in 1986. Since then the number of tests have grown rapidly. Today, APHIS has issued more than 323 permits and many of those permits are for multiple trials in multiple states. Pioneer has safely conducted more than 30 field trials in North America, with 5 different crop species and 17 different traits. These traits include virus resistances, herbicide resistances, visual markers, tissue specific gene promoters, inhibitors of gene expression, insect resistance, and nutritional modification.

Concerns continue to be expressed about the scientific validity of regulating transgenic plants as plant pests, and there are renewed questions about whether the regulations are process based, and therefore inconsistent with the 1992 policy of *federal oversight within scope of statutory authority: planned introductions of biotechnology products into the environment*. If APHIS were drafting rules today, I doubt they would use the same approach. However, in practical terms APHIS' regulation of small scale field trials has been successful. Even some public interest groups acknowledge that the system has merit.

The notification system, together with the provisions of the *Freedom of Information Act*, ensure all interest groups have access to information about proposed trials that would not otherwise be readily or publicly available. Without Part 340 of the *Plant Pest Act*, I venture to speculate that we would have seen a groundswell of public opinion resulting in a host of different states implementing their own regulations and permitting processes; possibly applying inconsistent standards that might not always be science-based or risk-focused. Clearly industry has found the APHIS system of oversight under the Act an asset in promoting public confidence and certifying that standards have been met during small scale field testing.

Several companies have been able to bring products close to commercialization, but obviously they cannot be sold on the open market if growers need a permit every time they plant a seed. Two companies, Calgene and Upjohn, have recently applied for interpretive rulings by APHIS that their transgenic tomato and squash plants do not pose a plant pest risk; therefore, will no longer be considered regulated articles requiring permits. In both cases, the petitions are less than 50 pages long -- shorter than our first field trial application! If these petitions are approved, we will be well on the way to commercializing of the first biotechnology plant products.

If APHIS finds that such plants no longer represent regulated articles, they will, in effect, be certifying these particular combinations of vectors, control sequences and genes do not present a plant pest risk. It is unlikely that decisions on these applications will have broader utility for plants bearing similar sequences developed by researchers or other companies. In other words, a finding of no plant pest risk for Calgene's tomatoes will not exempt disarmed *Agrobacterium* vectors from oversight, but could be viewed as a certificate or license to commercialize that specific product, subject to compliance with existing pesticide and food safety statutes.

The Regulation's Impact on Academic Research

What about the impact of regulations on the academic community? I think it is interesting to look at my colleagues in England and see what they are doing today. In many cases, the regulatory debate has had a direct effect on broadening their academic horizons and sources of funding. Some are now engaged in multi-disciplinary research, together with ecologists, population biologists and statisticians, investigating the environmental impact of small scale field trials and the potential impact of genetically modified disease resistant products. Such studies have been funded by organizations such as the European Community or industry consortia. One or two of my colleagues have served as scientific advisors to government bodies engaged in drawing up guidelines for the regulation of biotechnology products. Possible commercial applications of their research have become much more important. Some now hold patents and many are working on projects that are partially or completely funded by industry, an experience that has also tended to broaden their appreciation of regulatory issues associated with biotechnology. Closer ties with industry has also served to blur the distinction, at least in the public's mind, between academic and commercial research. Exactly the same trends are evident in the United States.

Despite these specific impacts on particular researchers, from a more general perspective, it is important to note that of the current total of 323 field trial permits issued by APHIS, only 59 were issued to university or ARS scientists. Is this evidence that regulations are having a negative impact on public research? Cases exist where university scientists have successfully obtained a permit after filing an application on eight pages torn from his laboratory notebook. However, there is still a perception in the academic community that getting a permit is more complicated than it really is. Nonetheless, I do not believe that is the major explanation for this disparity. According to a survey of the impact of regulations on research that was conducted in 1990, it is industry that feels more constrained by regulations. Of 355 respondents to the question, "Have you ever been discouraged from conducting field tests with genetically modified organisms?," 16% of public sector researchers replied yes. Whereas 23% of private sector respondents felt regulations had inhibited their field research.

The distribution of permits seems to reflect different research priorities and patterns of investment. A large portion of academic research in plant biology and pathology has been in the area of fundamental molecular biology an area where the largest investment of public funds in biotechnology has taken place. The researcher who is investigating a new molecular mechanism of virus resistance, for example, does not need to conduct multiple field trials in order to validate his hypothesis and publish the results. Instead, a series of well designed growth chamber or glasshouse experiments will often suffice.

By contrast, a commercial company is ultimately interested in how those plants perform in an agricultural context; hence, the priority is to test new lines in multiple field locations where the stability of that resistance can be assessed. It is not surprising to see the commercial sector conducting far more trials. In fact, we can expect to see this trend become even more apparent in the coming years.

The Cost of APHIS Regulation

A very real impact for all of us is the cost of regulatory compliance. The expense of making applications was a major disincentive in the early years. Field test applications were usually complex scientific documents supported by extensive molecular and genetic data and a stack of references several inches high. I worked on Pioneer's first field trial application for three months. There was a strong feeling we needed to show publicly how thoroughly our material was characterized and that we had attempted to consider all environmental risks, notwithstanding APHIS own environmental assessment. I recently calculated that the first application cost over $10,000 to prepare -- twice APHIS' estimate of $5,000. I know that the Agency would claim we submitted much more data than was required. Like most other companies and many universities, we also conducted a national and local public information program that cost a lot more than $10,000.

Even then we still couldn't control nature -- heavy fall rains water logged the site of our trial, and the hard winter of 1989 killed all but a few of the virus resistant alfalfa plants. Three years later, the weather is no more cooperative but, it takes us less than a week to prepare a completely new permit application and costs around $500 -- well below the Agency's estimate. Of course even that figure is still far too high and must be lowered.

What about the cost of additional safety precautions. I remember visiting Dr. Thornberg's field test of transgenic tobacco at Iowa State. This was the first field test to be conducted by a public institution. Although tobacco plants are few and far between in Iowa, the site was isolated from other field crops, bordered by a zone of barren ground and surrounded by a purpose built fence. Dr. Thornberg had to construct a special closed container covered with biohazard labels to transport his plants from the lab to the field - a distance of a couple of miles. The plants were not even allowed to flower.

Today, Pioneer has transgenic corn growing at several different locations in Iowa, Nebraska -- the heart of the corn belt -- and Tennessee, and at our winter nursery in Hawaii. Transgenic plants grow alongside non-transgenic material as regular entries in our research nurseries. Standard industry practices are used to contain pollen, and crosses are made in the field. There are no special fences, barriers or border rows. Apart from the requirement to destroy all grain harvested from the experimental material, it is almost business as usual for these small scale trials.

For the most part, we have not felt unduly constrained by the conditions imposed upon us by APHIS. There have been attempts to impose what we saw as unnecessary requirements, but the Agency has been most responsive to our concerns. Where we have been able to show why those particular precautions were unnecessary, the conditions of the permit have been changed.

An Assessment of APHIS

So are we satisfied with the APHIS system? Certainly not. The regulation that has arguably done the most to promote the development of US agricultural biotechnology also has the potential to bring the whole thing to a grinding halt! APHIS reviewers struggled to meet the demands placed upon them by the number of permit applications submitted for this year. Most applications took the full 120 days to review -- a few even longer. We had field researchers virtually sitting on their planters waiting for permits to be faxed to them. If APHIS thought this year was bad -- let me tell you how bad it can get!

Part of the development process of a new corn hybrid is wide area testing. During this phase, the hybrid is grown in small plots at multiple locations across the corn belt to confirm that it performs in a predictable manner. For example, in 1988 we commenced wide area testing of Pioneer® brand hybrid 3394 at 18 locations. The next year it was tested at 73 locations and the next year at 565 locations. In the year before it was placed on the market, the hybrid was tested at a total of 2,562 locations. In the near future, we may begin wide area testing of transgenic corn hybrids and soybean varieties. I do not want to have to file permit applications for all those tests and I am certain APHIS doesn't want to review them all -- especially as they are all essentially the same.

Clearly, USDA is not going to receive additional resources to administer the *Plant Pest Act*. If the system does not change, then next year has the potential to be a disaster for us all. We are trying to beat the rush by preparing permits for next May, but this is not always practical. Many factors can change in 9 months, besides which, a 120 day review period is totally incompatible with operating an efficient corn or soybean breeding program. Especially breeding programs that use winter nurseries, where the turnaround between harvest in Hawaii or Mexico and planting the midwest may be a couple of weeks. APHIS is aware of the problem and is proposing a move toward self-regulation, based on a notification procedure -- the same move industry proposed to Congress two years ago.

APHIS has held a series of conferences to address safe field testing of different crops. As a result of information presented at those conferences by academic and industry experts, supported by a 100% record of safety, APHIS is proposing a performance based standard for field trials with certain crops and certain genes. This proposal was recently presented to the ABRAC where it reportedly received qualified support.

For crops such as corn, potatoes, tomatoes, and soybeans that have little or no potential to outcross to weeds and that are transformed with genes that we are confident do not pose a plant pest risk (for example, coat protein genes from indigenous plant viruses), an applicant would simply notify the agency of their intention to plant a test and certify that they will maintain agency approved performance standards for the crop. These would be based on good agricultural practices and isolation distances for certified seed production. Applicants would also have to acknowledge the right of Agency officials to inspect the trial and give an assurance that the crop will not be used for food or feed. We estimate that the cost of such an application would be minimal -- perhaps just the price of a postage stamp or a FAX.

A number of states appear prepared to endorse the performance standards and would accept this proposal subject to continued provision for notification of planned releases. From what we have seen of the proposals, they could form the basis of a workable system and that would represent a logical step toward eventual deregulation of certain groups of organisms.

I believe that a lot of the frustration expressed with the APHIS oversight under the *Plant Pest Act* is actually generated by the requirement to obtain permits for interstate movement of genetically engineered materials. Amongst our own researchers, the most frequently heard complaint is that it takes 60 days, occasionally longer, to obtain a permit to move something as innocuous as one-half dozen small pieces of leaf from a transgenic soybean plant between Nebraska and Iowa. One member of our staff spends three quarters of her time filing applications for interstate movement permits.

I refuse to believe that a small piece of corn leaf or a delicate plant tissue culture is going to burst out of its Federal Express package and establish itself as a vigorous and invasive plant pest by the side of the road. If the field testing system can move to notification and performance based standards, so can the interstate movement regulations. I believe these changes can be made without the necessity to exempt broad groups of organisms from the Act; indeed, I am sure this is APHIS intention. Performance based standards for contained research, supported by notification and provisions for spot checks to verify compliance, should be sufficient to satisfy the quarantine requirements of most states.

All these proposals will certainly make my life much easier, but are we missing an opportunity to move further? The *Plant Pest Act* contains provisions for creating exemptions based on the lack of any evidence of plant pest risk. Why not immediately create broad classes of exemption for such things as disarmed Ti-plasmids or the cauliflower mosaic virus 35S promoter? Perhaps we should throw out Part 340 altogether, on the grounds that it is process based. That would appear to free public sector researchers of any responsibility under federal regulations. On the contrary, for the reasons I will discuss, such moves would have a negative impact on biotechnology research in both the public and private sectors.

Firstly, in the absence of APHIS oversight, environmental releases might be regulated by the Environmental Protection Agency under either FIFRA or the *Toxic Substances Control Act* or [ToSCA], which is designed to be gap filling legislation. Although ToSCA presently has an exemption for non-commercial research with chemicals, the Agency's 1991 draft ToSCA regulations for microbes appear to take the stance that since genetically modified organisms are capable of independent reproduction, they pose the same risk to the environment, irrespective of their intended use.

Secondly, if no Federal agency has authority over field releases, then presumably any review authority would devolve to the level of an internal group such as the Institutional Biosafety Committee or an equivalent organization with agricultural experience, perhaps operating under the 1991 ABRAC *Proposed USDA guidelines for research involving the planned introduction into the environment of organisms with deliberately modified hereditary traits*. For those receiving federal research funds, these informal oversight mechanisms may not comply with the requirements of NEPA. Furthermore, the costs of NEPA compliance would now fall on the research institution itself instead of APHIS.

Thirdly, and most importantly, in the absence of Part 340, there would no longer be a mandate for public notification of proposed tests. Almost every report you read (for example, Dr. Hoban's recently published survey of *Consumer attitudes about the use of biotechnology in agriculture and food production* or the latest Congressional Office of Technology Assessment Report, *A New Technological Era for American Agriculture*) make the same point -- public acceptance is key to the future of agricultural biotechnology.

I should emphasize that in referring to the public, I include those groups that claim to represent the public, the special interest groups, and the popular media. If the public does not accept the products, then I and my colleagues will be out of a job and private and public funding for a broad range of biotechnology research will dry up. Now more than ever, we need to be promoting public confidence by informing consumers about safe testing and practical applications of biotechnology research in agriculture.

In the United States, activists may no longer dig up transgenic plants and the press no longer wants to run stories about safe field tests. However, that does not mean that people now have confidence in the technology or trust the technologists. I predict that creating broad exemptions from the Act or doing away with Part 340 will ultimately result in the majority of states implementing their own regulations, as they seek to protect their citizens from the perceived dangers of unregulated field research with transgenic organisms. The outcome of such local attempts to regulate biotechnology are notoriously uncertain. Moreover, regulation may not stop at the state level. Local ordinances are also likely to be enacted, paralleling the host of local restrictions on pesticide use that followed the supreme court decision in *Wisconsin Public Intervener versus Mortier.*

I do not subscribe to the theory that states are strapped for cash and cannot afford to implement their own regulations. The State of Minnesota simply imposed user fees starting at $5,000 per application with no cap. As far as I am aware, there is no waiver for publicly funded institutions. For a company that has research stations located across the United States, the prospect of having to deal with 20 or 30 different sets of state regulations is a daunting one, especially if those states impose user fees to cover their administrative costs.

Even for a university, the impact may be considerable. An institution may be well respected as a seat of learning, be on excellent terms with the governor, and have a good working relationship with the Department of Agriculture; however, when the governor is thrown out, and a public interest group moves in, and the Department of Health or the Department of Environment is assigned the responsibility for approving field trials, things may look very different.

Some states such as Florida, Hawaii, Washington, and Wisconsin have systems which already require notification by the applicant of plans to conduct field tests with genetically engineered organisms. Such regulations often defer to the federal regulatory process if a significant review of the proposed trial has been made by a government agency. In the absence of Part 340, these local regulations would likely be expanded to include a formal state review with public comment period.

Field trials have been conducted in states like North Carolina and Minnesota that have their own regulatory review and permitting systems. However, those systems are constructed to be consistent, at a minimum, with the 120 day review period stipulated by APHIS. Even if the Act goes away, the state review will not get any quicker and might be slower. Under certain circumstances, the Minnesota law could slow the application process to the point where researchers would only be able to conduct one test every other growing season. If the regulatory situation in a particular state became untenable, then a company could relocate our operations elsewhere. A university does not have that option.

Earlier I spoke of the 17 different traits that we have field tested. I think it is significant to note that five of those traits were developed in collaboration with university or ARS researchers. Successful technology transfer is promoted by a stable and predictable regulatory climate. In the near future, we are likely to see many more partnerships between industry and academia that will go beyond the lab and into the field.

Industry and academia, together with the regulators, share the responsibility for the past history of safe field testing and good public relations that has made the United States the leader in agricultural biotechnology research, particularly field research. We all share a common need for flexible regulatory systems that allow scientists to conduct safe research in a timely manner, under appropriate and cost effective oversight, while continuing to promote public confidence.

Chapter 11

Risk and Reason: An Assessment of APHIS

Susanne L. Huttner
University of California, Los Angeles

Biotechnology has been heralded as an important source of economic development for the 21st century. The rate of development of new knowledge and tools has been remarkable -- especially in biomedicine, where clinical scientists have already proceeded with the first trials of human gene therapies. Exciting benchmarks have also been achieved in agricultural sciences. Overall, however, the rate of development has been substantially slower in agriculture than in biomedicine. Perhaps more so than for other technologies, the rate and directions of seminal biotechnology developments have been strongly affected by governmental actions. Certainly to an extent previously unseen in life science research, federal policies have been determining how some of the very best investigative tools are used in research and training. The difference between use of biotechnology in the biomedical and agricultural sectors can be traced in part to the federal science and technology policy development process. The worthy goal of the coordinated federal approach to biotechnology policy -- to limit potential hazards while fostering research and economic development -- has proven difficult to attain. There is a critical need to reassess federal policy, especially as it affects basic research. There is a need to achieve greater balance between governmental regulation and incentives, so that public benefit from federally funded research in agricultural biotechnology is not deferred.

Much of the progress in biotechnology is the fruit of academic research. Indeed, what we call the "new biotechnology" derives from seminal developments in molecular biology in academic laboratories.

These developments have greatly extended and refined our abilities to manipulate genes. The scientific substrate that spawned the new techniques was provided by decades of public funding of basic research in the life sciences, primarily by the National Institutes of Health (NIH). Public support has enabled universities to supply the knowledge, techniques, and human resources that are the lifeblood of the biotechnology and related industries.

The new genetic technologies are universal, insofar as the biochemical basis of the genetic code is highly conserved. The opportunities and needs for applying the new tools are certainly no less great in agriculture than in biomedicine. The reason for the slower pace in agriculture can be traced, in part, to the academic, basic research community and to at least two federal policy factors that affect academic research: research funding and regulation.

Biotechnology in Agriculture vs. Biomedical Research

Biomedical research is better funded. A recent report by Goodman points out that, on average, competitive grants for plant biology have been 33% smaller and 25-50% shorter in duration than biomedical research grants. As a result, plant scientists must truncate their research goals and spend more time writing grants and renewals. This restrictive funding environment and its impact have been the subject of several national reports on U.S. competitiveness in agriculture, Hess and Hullar. These reports consistently urge a national agricultural policy emphasizing greater investment in broad-based, peer reviewed research and training.

Regulatory policies have played an equally important role in the development of agricultural biotechnology. Regulation of biomedical applications has been more reasonable as judged by regulatory policies for similar technologies and products. In contrast, entirely new regulatory requirements have been created just for agricultural biotechnology research and products. For the first time in the history of plant breeding, the federal government is scrutinizing early stages of crop research.

In the mid-1970's, all recombinant DNA (rDNA) research at federally funded institutions was strictly overseen by the NIH (1986 and 1991), but by 1982 the vast majority of laboratory experiments was exempted. New biopharmaceutical products moved towards commercialization. The Food and Drug Administration (FDA) announced as early as 1980 that it would treat new biotechnology products the same as other products -- without any special regulations, procedures, or requirements. Encouraged, investors and companies fueled robust activity in pharmaceuticals, with large numbers of biotechnology-derived products moving through FDA review.

The much slower evolution of agricultural regulation can be characterized as influenced by a general reluctance by regulators to follow FDA's lead. This has resulted in greater government intervention in basic plant research. The United States is not unique in its approach to agricultural biotechnology. New regulatory proposals and requirements have been created by governments around the world. In Europe, the worst case scenario is being played out in the most restrictive and scientifically irrational regulatory schemes. Scientists and industry are reportedly fleeing Germany, Kahn and Abbott.

These regulatory schemes conflict with broad scientific consensus that rDNA techniques and rDNA-modified organisms are not inherently dangerous or unpredictable. That consensus prompted the NIH to relax oversight of laboratory research. It also supported a 1989 National Research Council report that concluded (see Burris, Kelman, and Miller et al.):

> "Crops modified by molecular methods should pose risks no different from those modified by classical genetic methods for similar traits. As the molecular methods are more specific, users of the methods will be more certain about the traits they introduce into the plants."

From millions of laboratory experiments and hundreds of field tests, the scientific community has found that the precision of rDNA techniques actually enhances determinations of safety and risk. This confidence is not, however, reflected in federal regulations on field research. Both the Environmental Protection Agency (EPA) and the U.S. Department of Agriculture (USDA) have developed and implemented special regulatory requirements that selectively burden biotechnology research unbalanced by any incremental safety enhancement.

At the time these requirements were developed there was significant concern and uncertainty about public acceptance of field research. The specter of the "ice minus" bacterium controversy prompted many scientists to support scientifically flawed but workable schemes as a buffer against public interference in research. What has resulted, however, is a myopic focus on imagined risk scenarios to the near exclusion of actual risk data or conceivable benefits. It has lead to a serious under-representation of academic scientists in field research, showing the fragility of academic research incentives.

The hard lesson learned from efforts to correct these regulatory schemes is that bad science makes bad public policy -- and it is difficult to change the course of policy once it is set. It provides an interesting case study on the role of the interface between researchers, regulators, and activist members of the public in determining the course of science and technology development on a national level. For agricultural biotechnology, pressures applied at this interface have effectively taken the "science" out of science policy and replaced it with interpretations of safety and risk unaffected by empirical evidence.

A Reconstruction of the Rationale for Special Treatment of Agriculture

The government position on agricultural biotechnology is typically based on four arguments: (1) the regulations are needed for risk management; (2) they are good for commercialization; (3) the public wants them; and (4) the states will regulate if the federal government does not. It is worth exploring the rationales behind these arguments to gain an appreciation of the factors affecting national policies. The USDA Animal and Plant Health Inspection Service (USDA-APHIS) regulations (see USDA 1987) serve as a useful case in point.

Rationale: rDNA-modified Organisms are Sufficiently "New" to Warrant Case-by-Case Regulation

According to this rationale, the 350 field research permits issued by USDA-APHIS to date may seem a significant milestone -- if they were needed. Viewed in an historical context, they are more a millstone than a milestone because the majority involved plants posing negligible or low risk. Research with these plants arguably could have been conducted using standard good field testing practices without government intervention.

APHIS has interpreted the Plant Pest Act and the National Environmental Policy Act to require environmental assessments for the majority of rDNA-modified plants proposed for field tests. RDNA-modified plants have been regulated as if they were plant pests (Gasser and Fraley). They contain DNA sequences derived from plant pests -- typically disarmed Ti plasmids from Agrobacterium tumefaciens or 35S promoter from cauliflower mosaic virus. These nucleotide sequences ensure the efficient delivery and expression of introduced genes. They are part of a widely used system for introducing genes into plants. We know a great deal about the pathogenic systems from which they were derived. There is no reasonable scenario by which they could convert a crop plant into a pathogen (Huttner et al.). Therefore, the introduction of these DNA sequences should not trigger USDA-APHIS regulation under the Plant Pest Act. They didn't before 1987.

In 1987, USDA-APHIS created the new regulations (specifically for genetically engineered organisms) that subject certain field research to environmental assessment. Despite evidence from laboratory and greenhouse experiments on the absence of plant pest risk, the presence in a plant of material from a pest is sufficient to trigger regulation as a "regulated article." This effectively disconnects the Plant Pest Act from a clear definition of what constitutes a plant pest. It does, however, create a mechanism for subjecting most rDNA-modified plants to government scrutiny.

The view that these plants warrant special evaluation is simply not supportable. They are created from common, well characterized crop species -- like corn, soybean, walnuts, and tomatoes -- into which are introduced one or a few genes. As noted in Table 1, most of the introduced genes confer agronomically familiar traits such as insect or disease resistance, herbicide tolerance, improved nutrient composition, or improved post-harvest characteristics (Gasser et al.). These are the same goals sought by plant breeders for hundreds of years. In the new biotechnology plants, we couple our extensive knowledge of the agronomic history of the host crop with our knowledge of the specific genetic change. That is not to say that these plants pose no risks in field trials. They pose no unusual or extraordinary risks that warrant special government scrutiny.

Table 1

A Partial Listing of Introduced Traits

Pathogen resistance
- viral coat protein (potato virus x and y, potato leaf roll virus, tobacco mosaic virus, beet curly top virus, tobacco etch virus, tomato mosaic virus, tomato yellow leaf curl virus, cucumber mosaic virus, alfalfa mosaic virus)
- chitinase
- cecropin

Insect resistance
- Bacillus thuringiensis endotoxins

Herbicide tolerance
- nitrilase (bromoxynil)
- EPSPS (glyphosate)
- acetolactate synthase (sulfonylurea)

Product composition
- storage protein content
- dry matter (starch)
- oil modifications
- lysine concentration

Postharvest characteristics
- ripening or softening (antisense polygalacturonase, antisense pectin methylesterase)

Rationale: Government Regulation of Field Experiments is Necessary to Evaluate and Mitigate Risks

The USDA-APHIS approach asserts that governmental assessments are needed to judge potential plant pest risks. Is government regulation really needed? Is there any reason to believe that traditional breeding practices won't control risks of rDNA-modified plants just as they have for conventionally bred plants? Thousands of new breeding lines and cultivars have been introduced over the years by U.S. plant breeders. Many have been created by wide crosses with wild, weedy, and sometimes toxic relatives (Goodman et al.). Before 1987, USDA-APHIS did not conduct environmental assessments of initial field trials of plants with new genetic modifications.

There is, of course, always the possibility of surprise in both the primary and secondary effects of genetic modifications. Plant breeders traditionally have been concerned that new traits may have undesirable effects. In fact, the consequences of such unexpected effects strongly discourage frivolous releases. Disincentives work. Academia and industry have developed extensive procedures to identify and mitigate potential problems.

The traditional crop development plan is conceptually simple (Table 2): A gene resource is identified; the germplasm is introduced into a population of plants; new breeding lines with desirable combinations of traits are selected; they are evaluated for field performance, consumer acceptability, and safety; and then they are either discarded, released as cultivars for general use, or used in further breeding research. Breeders address the same basic safety concerns as USDA-APHIS addresses, including potential genetic exchange with wild or weedy cross breeding relatives. Whether pollen or seeds are moving into or out of a test population, the concerns and containment procedures are essentially the same.

Table 2

Crop Breeding Plan

1. Identify gene resource for trait(s) of interest.
2. Introduce germplasm into a plant population.
3. Select progeny with desirable trait combinations.
4. Evaluate selected plants for:
   a) Agronomic performance
      - growth
      - yield
      - susceptibility to insects and pathogens
      - pest control requirements
      - potential genetic exchange with cross-breeding relatives
      - relative viability and longevity of superior characteristics

   b) Safety
      - environmental
      - health

   c) Consumer acceptability
5. Release superior plants for general use. Destroy or return to research any unacceptable plants.

The safety record with conventional cultivars shows that, to the extent that new traits can be targeted for evaluation, breeders' assessments should be able to identify problems. In fact, new traits can be better targeted with rDNA methods than with classical breeding. As the 1989 National Research Council report emphasized, rDNA methods actually strengthen the breeder's ability to evaluate safety (Hardy). Taken together, extensive experience with rDNA, with conventional crops, and with researcher-driven breeding practices dictates a more moderate approach to oversight of the new biotech plants. Unless USDA-APHIS has evidence of serious unanticipated risks, traditional breeding practices should identify and mitigate problems for the vast majority of new plants. The scarce resources presently devoted to USDA-APHIS assessments should be limited to those plants that present potential plant pest risks, judged by consideration of the introduced traits, of the characteristics of the host plant, and of the proposed field test site.

Rationale: Government Regulation is Good for Commercialization

According to McCammon and Medley, the APHIS approach considers that the potential risks associated with rDNA-modified plants do not necessarily fall into familiar categories, such as pathogens, carcinogens, or chemical pollutants. The required environmental assessments are viewed as an integral part of the commercialization process, providing a mechanism for identifying and addressing emerging concerns. In that sense, APHIS provides verification that supports commercial development. It is not necessary that the genetic modifications involved pose any plant pest risk; regulation can be triggered by perceptions of plant pest risk.

The problem inherent in this approach is that it offers no means of distinguishing between what level of government action is appropriate for commercial research and what is appropriate for basic, academic research. The policy fails to take into account that unnecessary governmental requirements create a burden on academic research Ratner and Rabino). This, in turn, creates a disincentive for the transfer of research developments from university to industry. Historically, researchers at land grant universities have played a significant role in the development, evaluation, and distribution of new crop cultivars and germplasm -- both as a public service and as a responsibility. Major U.S. crops are either direct products of publicly-supported research or have in their breeding history germplasm derived from university research. University field trials provide necessary validation of performance and safety. Where over-regulation slows or restricts these trials it can impede commercialization of valuable new products.

Researchers in universities and industry may have different perceptions of regulations. They undoubtedly have different thresholds and tolerances for regulatory burdens. Industry factors regulatory requirements into development costs and, eventually, into the pricing for commercial products. A government imprimatur may even add value -- impressing investors and consumers, and possibly affecting liability should a mishap occur.

Academics have no mechanism for recovering the costs taken from already strained research budgets. They have no regulatory affairs staff. Even with the support of USDA-APHIS staff, an application typically requires at least several days preparation; some first-time applicants have invested as much as several weeks, and others have abandoned projects altogether. Moreover, the cost of a single field test is typically 10 times greater than that for a plant produced by conventional techniques.

It is not surprising that fewer than 15% of all USDA-APHIS permits have been submitted by academic researchers, individually or in collaboration with industry applications. An undeniable administrative and financial burden weighs on every permit application. It translates into under-representation of academic scientists and a delay or deferral of important preliminary field trial validations of new plant varieties.

Rationale: The Public Demands Regulation

The public does want protection from risks. In fact, some in our society expect and demand zero risk. However, the public also demands accurate information on risk. They expect not to be mislead by government. When government requires special evaluation where it is not needed, it creates an aura of risk, fostering unwarranted public apprehension. The false message of risk eventually may undermine public acceptance of research and new products. Indeed, where technological improvements are perceived by the public as marginal -- as most agricultural biotechnology developments may seem to consumers -- even small risks will likely be judged as unacceptable.

Some regulators assert the opposite -- that the public is reassured by regulation. However politically expedient that may appear in the short run, in the long run the public interest suffers when government unnecessarily impedes publicly-funded research that is conducted in a responsible manner. Government should regulate risks, not misperceptions. We need to explain why scientists judge the risks associated with rDNA-modified organisms to be commonplace and to fall into long-established categories. We should also acknowledge that no test -- developed by USDA or anyone else -- will provide assurance that organisms will not spontaneously mutate, exchange genetic material with other organisms, modify their metabolism to accommodate changes within their niche, develop drug or chemical resistance, or proliferate into new environments. No government policy can ensure zero risk.

We must move away from policies guided by government perceptions of public perceptions of risk, and focus policy on substantive concerns about risks and address public perceptions with appropriate public information policy. We must move towards a policy of communication and, where misinformation or misperceptions are found, of public education. Factual communication and ongoing dialogue is essential to ensuring public confidence. A new partnership between government and academic institutions is called for to support public understanding of emerging issues and to foster more effective public participation in government decisions.

Rational: If Federal Agencies Don't Regulate the States Will

In 1990, a survey (Huttner et al.) of state regulators showed that most states were not considering new regulations and some were simply not interested in agricultural biotechnology issues. A few states have enacted stringent regulations, notably North Carolina and Minnesota. However, more have considered and rejected such proposals -- including California, Florida, Hawaii, Kentucky, Louisiana, and New Jersey. As early as 1985, agencies in California and Florida decided biotechnology research and products were not so unique that existing authorities and procedures would be inadequate. That was before USDA-APHIS implemented biotechnology regulations and just after California had reluctantly hosted the "ice-minus" bacterium controversy. Unfortunately, a decision not to promulgate new regulations gets little media coverage, while a state even considering the development of new rules is widely reported. As a result, there is a general misperception that the states are poised to write 50 versions of field research regulations should USDA-APHIS relax its rules.

While many exercise authority over agricultural activities, states generally view regulation of agriculture as necessarily requiring a partnership with federal government. Thus, any changes in federal policy must be communicated in a straightforward and timely manner. USDA-APHIS has made a serious effort to build effective lines of communication with state officials. This has fostered a sense of trust and confidence that will enable them to explain the needed changes. That relationship can also be used to extend USDA's reach to state legislatures and local communities interested in agricultural biotechnology regulations. State and local governments have an historical role in oversight of agricultural activities. Their analyses of biotechnology research deserve and, indeed, require the assistance and support of the academic and industrial research communities.

Summary and Conclusions

I believe this decade will be marked by important successes in agricultural biotechnology -- but only if funding for basic research and training improves and if the regulatory system becomes reasonable and unambiguous. Biotechnology is projected to have tremendous impact on national and international economies. It provides valuable new tools for the agricultural toolbox -- helping farmers respond to the world's growing demand for food from scarce natural resources -- demand created by a world population expected to reach 8 billion by the year 2030. To meet these needs and to compete in world markets, we must use the best tools and adequately support research.

However, adequate funding does not guarantee success. As Charles Arntzen has stressed: "A pipeline filled with commercial rDNA-modified plant products does not necessarily mean that consumers will rapidly and efficiently benefit from scientists' discoveries ... U.S. government regulations ... remain complicated by bureaucratic detail that stifles the research which provides the basis for new product development." Researchers in academia and industry have worked with USDA and EPA. It is now reasonable to expect either that the regulatory burden will be eased or that the agencies will present evidence on any discovered risks so that they may be studied. After five years and more than 600 carefully studied field trials worldwide (McKenzie et al. and Landsmann) USDA-APHIS should be able to report what specific risks warrant government intervention in research.

Until that question is addressed, the current regulatory scheme will continue unnecessarily to burden field trials, to discourage academic researchers from using the best new techniques, and to slow agricultural innovation. Moreover, it will continue to perpetuate the public mythology of technology and risk. Good science and good sense both dictate the realignment of USDA-APHIS and EPA regulations to impose only that regulatory burden that is justified by and commensurate with risk. Then, U.S. agricultural biotechnology will have full opportunity to demonstrate its potential to serve the nation's agricultural, environmental, and economic development needs.

References

Abbott, A. 1992. "German Institutes, Industry Seek to Ease Genetic Engineering Laws." *Nature*, 359:93.

Arntzen, C. J. 1992. "Regulation of Transgenic Plants." *Science*, 257:1327.

Burris, R. H. (Chairman). 1989. "Field Testing Genetically Modified Organisms: Framework for Decisions." Committee on Scientific Evaluation of the Introduction of Genetically Modified Microorganisms and Plants into the Environment, National Research Council. National Academy Press, Washington, DC, p. 3.

Gasser, C. S., and Fraley, R. T. 1989. "Genetically Engineering Plants for Crop Improvement." *Science*, 244:1293-1299.

Goodman, R. M. (Chairman). 1992. "Plant Biology Research and Training for the 21st Century." Committee on an Examination of Plant Science Research Programs in the United States, National Research Council. National Academy Press, Washington, DC.

Goodman, R. M.; H. Hauptli; A. Crossway; V. C. Knauf. 1987. "Gene Transfer in Crop Improvement." *Science*, 236:48-54.

Hardy, R. 1987. "Regulatory Considerations: Genetically Engineered Plants." Boyce Thompson Institute for Plant Research Workshop. Center for Science Information, San Francisco.

Hess, C. E. (Chairman). 1987. "Agricultural Biotechnology." Committee on a National Strategy for Biotechnology in Agriculture, National Research Council. National Academy Press, Washington, DC.

Hullar, T. L. (Chairman). 1989. "Investing in Research." Board on Agriculture, National Research Council. National Academy Press, Washington, DC.

Huttner, S. L.; C. Arntzen; R. Beachy; G. Bruening; L. DeFrancesco; E. Nester; C. Qualset; and A. Vidaver. 1992. "Revising Oversight of Genetically Modified Plants." *Bio/Technology*, 10(9):967-971.

Huttner, S. L., and Smith, R. L. 1990. "State Case Histories on Biotechnology Oversight." University of California Systemwide Biotechnology Research and Education Program, Los Angeles.

Kahn, P. 1992. "Germany's Gene Law Begins to Bite." *Science*, 256:524-526.

Kelman, A. (Chairman). 1987. "Introduction of Recombinant DNA-Engineered Organisms into the Environment: Key Issues." Committee on the Introduction of Genetically Engineered Organisms into the Environment, National Research Council. National Academy Press, Washington, DC.

Landsmann, J. 1992. "The Biosafety Results of Field Tests of Genetically Modified Plants and Microorganisms." Rapporteur's report on the Second International Symposium. Goslar, Germany.

MacKenzie, D. R., and S. C. Henry. 1990. "The Biosafety Results of Field Tests of Genetically Modified Plants and Microorganisms." Rapporteur's Report of the First international Symposium. Agricultural Research Institute, Bethesda (1990).

McCammon, S., and T. L. Medley. 1992. "Commercialization review process for plant varieties developed through biotechnology." In Rapporteur's Report of the Second International Symposium on The Biosafety Results of Field Tests of Genetically Modified Plants and Microorganisms, Goslar, Germany.

Miller, H. I.; R. H. Burris; A. K. Vidaver; and N. A. Wivel. 1990. "Risk-based Oversight of Experiments in the Environment." *Science,* 250:490-491.

National Institutes of Health. 1986 and 1991. "Guidelines for Research Involving Recombinant DNA Molecules." Federal Register, 51:16958 and Federal Register, 56:51784.

Rabino, I. 1991. "The Impact of Activist pressures on Recombinant DNA Research." Science, Technology, & Human Values 16:70-87.

Ratner, M. 1990. "Survey and Opinions: Barriers to Field-Testing Genetically Modified Organisms." *Bio/Technology*, 8:196-198.

U.S. Department of Agriculture. 1987. "7 CFR Parts 330 and 340, Plant Pests; Introduction of Genetically Engineered Organisms or Products; Final Rule." Animal and Plant Health Inspection Service. Federal Register, 52(115):22892-22915.

Intellectual Property Rights Law: An Incentive Based Approach to Managing Agricultural Research

Chapter 12

Intellectual Property Laws and Their Impact on Agricultural Research

William H. Duffey
Monsanto Company

Most readers hardly need to be reminded of the critical importance of agricultural research to the welfare of mankind worldwide. That fact seems axiomatic to all of us. This paper will consider our intellectual property laws in America and how they impact agricultural research. Intellectual property laws and agricultural research are very much intertwined. Indeed, they are inseparable. As a starting point, we can briefly examine the various intellectual property rights which are available for the products of agricultural research.

Intellectual Property Rights Available for Agricultural Products

Time would not permit an exhaustive examination of each body of intellectual property law which is applicable to agricultural products. In the U.S., there are four categories of intellectual property law which are of interest to those doing agricultural research. These categories are utility patents; plant patents; plant variety protection certificates; and trade secrets. Each of these categories is quite distinct from the others. The scope of protection varies considerably and the method of protection and method of enforcement can also vary widely.

(a) Utility Patents

Utility patents for plants derive from the general patent laws. Utility patents provide the broadest and most effective protection for most products of agricultural research. Herbicides, insecticides and fungicides; transgenic plants; plant cells; plant genes; plant DNA sequences; plant tissue cultures; transgenic seeds; plant varieties; host-vector organisms; and many other products of agricultural research are susceptible to protection under the general patent laws of the United States in the form of utility patents. In most cases, therefore, utility patents are the preferred intellectual property right for the agricultural researcher.

(b) Plant Patents

Plant patents should be thought of as much narrower in scope than the utility patents described above. Confusion often arises because an inventor may acquire more than one type of plant patent. Congress enacted the Plant Patent Act (PPA) in 1930 to allow protection for new and distinct a sexually propagated varieties other than tuber-propagated plants. The PPA was designed to encourage new plant variety development and to afford agriculture the benefits of the patent system. But protection under the PPA is for only a single variety and not a group of varieties having a common trait. Utility patents for plants, therefore, are very much broader in scope of protection than plant patents under the PPA.

(c) Plant Variety Protection Certificates

In 1970, Congress enacted the Plant Variety Protection Act (PVPA). PVPA was enacted to encourage the development of novel, sexually reproduced plants by providing an economic incentive for companies to undertake the costs and the risks inherent in producing new varieties and hybrids. PVPA protection extends only to a single variety and not to a group of varieties having a common trait. Once again, PVPA like PPA, affords a much narrower scope of protection to the proprietor than that afforded under utility patents. PPA and PVPA complement each other in providing protection for new varieties of plants.

(d) Trade Secrets

Trade secrets, in addition to the three categories recited above, are also an important form of protection for agricultural products and technology. Trade secrets are subject to State law whereas the first three categories are subject to Federal law. Trade secret rights can be protected in laboratories and factories where the movement of outsiders is confined and security is maintained. If a Trade secret is disclosed in a nonconfidential manner, it is lost forever. With secrecy a legal prerequisite to a Trade secret, it can often be difficult for a commercial enterprise to utilize trade secrecy as a form of intellectual property protection.

Utility Patents Play A Major Role in Agricultural Research in America

From the perspective of the agricultural industry in America, there can be absolutely no doubt that utility patents play a central --- indeed crucial --- role in the progress of research and development. The record book is replete with incontestable evidence that patents are an incentive for investment of high-risk capital in agricultural research and development. That evidence is irrefutable.

Considering the enormous time, money and effort which American companies invest in agricultural research and new product development, it is critical that they be able to protect their valuable products against imitation. In return for having disclosed the invention to the public, for the public to thereafter improve upon the subject matter, the U.S. Government grants to the inventor the right to exclude others from making, using or selling the subject matter of the patented invention for a limited period, i.e. 17 years. The patent grant may be thought of as contract between the inventor and the government. Limited exclusivity is granted to the inventor in return for his/her public disclosure of the technology. The period of exclusivity allows the inventor and the inventor's sponsor to recoup the enormous costs of research and development. Patents, therefore, are an indispensable vehicle for encouraging research and invention.

Consider for a moment the history of herbicide research in America where thousands and thousands of candidate chemical compounds are prepared through classical organic synthesis and then screened for herbicidal activity. Published data confirm that only one out of perhaps 10,000 screened chemical compounds will ultimately provide economic value to agriculture. Discovery, development and regulatory approval of a new agrichemical entity can often require 8 to 10 years of effort at a cost well in excess of $100 million. Like new pharmaceutical molecules, agrichemicals are subject to strict regulatory scrutiny by governmental agencies. The crucial need for strong utility patent protection is self-evident.

Protecting Products of Agricultural Biotechnology

The exciting new science of plant biotechnology offers unprecedented potential for the betterment of mankind worldwide. The social and economic potential of this plant revolution is staggering. Today's agricultural scientists have developed sophisticated techniques for genetically transforming plants to achieve insect resistance, frost resistance, drought resistance, viral resistance and herbicide resistance. Research in plant molecular biology tends to move slowly because the plant genome is complex and not well understood. Regulatory delays add further time and expense for the agricultural biotech entrepreneur.

Intellectual property protection, therefore, takes on enormous importance in this nascent science of plant molecular biology. Variety protection under PVPA is entirely inadequate for transgenic plants. Variety rights provide no protection for broad inventive concepts, nor do they provide protection for plant parts such as plant genes, fruit, cut flowers, etc. A single utility plant patent, however, can cover a multiplicity of varieties as well as parts of those varieties and methods of producing them, thereby giving the inventor the breadth of exclusivity he/she deserves.

The inadequacy of traditional plant variety rights cannot be over-emphasized when discussing agricultural biotech patent protection. Plant variety rights provide protection for only a single variety as described in the Plant Variety Protection Certificate. This is simply not enough protection for the plant molecular biologist.

New traits imparted to transgenic plants such as insect resistance and viral resistance are not confined to a single plant variety but can affect a host of varieties of e.g. soybeans, sugarbeets, etc. It is for this reason that generic patent protection rather than individual variety protection is crucial for sheltering these modern inventions arising from plant molecular biology. The science of plant molecular biology is profoundly different from classical plant breeding through selection of parent lines.

Economic Impact of Intellectual Property Protection on U.S. Agriculture

Intellectual property protection continues to influence the direction of seed, plant and agrichemical research. Intellectual property rights are critical to continued investment in plant variety development. Innovation must be protected and rewarded to realize a continuing flow of money into agricultural research and development.

Intellectual property protection, however, is not without its critics. Patent protection is especially singled out for attack from some quarters. Some individuals are concerned that increased patent activity results in the privatization of agriculture and has adverse consequences for small farmers. Supporters of patent protection, on the other hand, have a rebuttal. Instead of taking something away from society, the advocates say a patent grant adds something new to scientific knowledge. By sheer definition, if the inventive contribution were not new, it could not merit a patent in the first place. Furthermore, farmers are not obligated in any way to use a patented seed. This is entirely optional. Farmers may continue to use the seed varieties which they are currently growing. Thus, the farmer would be inclined to purchase a patented seed only when it was deemed economically advantageous.

Nevertheless, there will always be debate within the agricultural community on the merits and the demerits of intellectual property protection. Advocates of PVPA will argue that private investment in new plant varieties has increased since 1970 when PVPA was enacted. Others will argue that the rate of private research investment in plant breeding following the passage of PVPA is no different than that during the decade preceding 1970.

Some individuals argue that a noticeable slowing in the free exchange of germplasm has occurred because of increased patent activity. Evidence supporting this proposition, however, is far from persuasive. But no one disputes the notion that advances in plant breeding and plant biotechnology require a free-moving, international exchange of germplasm.

The challenge which we face, therefore, is to maintain a continual balance of both societal goals and economic goals as we administer intellectual property protection for plants in America.

Strategic Advice for Agricultural Researchers and Entrepreneurs in Global Markets

For protecting intellectual property rights growing out of agricultural research, America stands alone as the most hospitable venue in the entire world. Protection in America for agrichemicals, transgenic plants, transgenic seeds, plant varieties and parts of plants, is by far the finest system available anywhere. This not only applies to the various systems for obtaining intellectual property rights, but also to the crucial issue of enforcing those rights against unauthorized actions by others. From a commercial point of view, this has enormous importance.

As the science and the products of agriculture increasingly take on a global rather than a national perspective, the agricultural researcher and entrepreneur must inevitably give thought to the intellectual property laws of foreign countries. With respect to traditional agrichemicals, plant varieties and seeds, it can be said that the intellectual property laws of major industrialized countries outside the U.S. are generally acceptable. But much improvement needs to be accomplished in many developing nations in this regard. Ongoing negotiations are currently underway, albeit slowly, in the Uruguay Round of the General Agreement on Tariffs and Trade (GATT) with a view to imposing minimum intellectual property standards on all countries which are signatory to GATT. But these hoped for improvements will most assuredly come slowly. Nevertheless, agricultural researchers engaged in non-biotech based work can be reasonably comfortable with the global outlook for protection and enforcement of intellectual property rights.

Regrettably, however, the outlook is far less sanguine in the nascent science of agricultural biotechnology. The patent laws of many important countries, mainly in Europe, do not afford with certainty the proper generic patent protection to those who are investing so heavily in plant biotechnology.

Thus, the ag biotech entrepreneur, whether large or small, should become accustomed to the notion that intellectual property rights on products of genetic engineering will not be ideally protected in all countries of agronomic importance. One who aspires for participation in international agricultural markets must therefore examine each target country individually while designing a property protection strategy most appropriate for that venue.

The business manager, venture partner, research manager or other decision-making party should design an imaginative strategy which takes advantage of alternative forms of protection other than (or in addition to) patents. Such alternative forms would include breeders' rights, plant variety certificates and product registration. While falling short of the protective power of patents, they can nevertheless provide some measure of shelter.

Another rather obvious form of property protection as an alternative to patents, is the use of legal contracts to control the disposition and confidentiality of intellectual property rights as between the contracting parties. The contracting parties might stand in the posture of licensor and licensee. Or they could be venture partners in which one or more partners contributes proprietary technology to the venture. Or the contracting parties might be a molecular biology firm on the one hand and a seed company on the other wherein the biotech firm transfers valuable genes and vectors to the seed company pursuant to a joint research undertaking or perhaps a straight commercial venture. Indeed, the novel gene/vector system might even by the subject of a commercial lease arrangement. And the seed breeder will possess valuable proprietary rights of his own.

Still another exclusivity shroud for the ag biotech entrepreneur or seed company is the time-tested use of hybrid varieties. The use of hybrid varieties has traditionally served as an anticounterfeiting tool for the plant breeder. Hybrid plants are inherently protected from duplication because they do not reproduce completely and faithfully from seed. By keeping secret the identity of the parent lines from which the hybrid variety was bred, the proprietor forces the user to purchase new seed for each planting season. While this scheme may have appeal at first blush, it is an unnatural, slow and unsatisfactory substitute for generic patent protection in the context of ag biotech.

Yet another umbrella of exclusivity accruing to the ag biotech entrepreneur resides in the competitive lead time inherent in the length of the seed breeding cycle itself.

Happily, therefore, the ag biotech industrialist or venture partnership has more than one mechanism for sheltering the expensive intellectual property emerging from ag biotech R&D. But there are pitfalls ahead if the industrialist exports that property without regard to the legal climate in the receiving country. This caveat may seem terribly obvious. But one would be astounded at the property losses occurring in those countries which fail to provide patent and trade secret protection --- that refuse to enforce contracts --- that harbor blatant disrespect, often contempt --- for foreign owners of valuable intellectual property.

Key Intellectual Property Issue Facing Ag Researchers

Plant molecular biologists have made remarkable progress in recent years in the genetic transformation of seeds and plants. New biotech products for agriculture will require staggering R&D investments followed by clearance of stringent regulatory hurdles. Without patent protection and the exclusivity it affords, the ag biotech investor will struggle to recover his/her costs. Ag biotech patent protection in the European community is notably deficient. But there is documented evidence that European trade ministers are vitally concerned that Japan and America will surpass Europe in the quality and quantity of Ag biotech research unless the biotech climate in Europe improves.

With considerable disappointment we have witnessed our neighbors in Canada stepping forward to steadfastly renounce the mandatory patent protection of transgenic plants and seeds. Indeed, Canadian negotiators were in the vanguard in the ongoing GATT trade negotiations as they advocated the optional exclusion of higher life forms such as transgenic plants from patent protection. In the recently-announced North American Free Trade Agreement (NAFTA) covering Mexico, Canada and the U.S., the Canadian negotiators succeeded in once again providing optional exclusion of higher life forms from utility patent protection. For those American companies which are spending enormous sums of money in research on transgenic plants and seeds, this development is very disquieting. While U.S. patent standards will not be diminished because of NAFTA, the Canadian Government can (and probably will) continue to deny Canadian patents for transgenic plants and seeds.

Nevertheless, ag biotech researchers and entrepreneurs must continue their quest for new biotech products of agronomic value despite the absence of utility patent protection in many important countries. We in America are especially advantaged in having a property rights system which is the paragon of the world.

The powerful science of agricultural biotechnology offers too much hope for mankind to be allowed to languish because of a flawed global patent system.

Chapter 13

Intellectual Property Rights in the Public Interest

Richard K. Perrin
North Carolina State University

Intellectual property rights have to do with products of the human mind. They bestow upon their owner the right to exclude other persons from certain kinds of uses of the protected ideas. The very notion of such rights is itself a product of the human mind, created to serve the public interest rather than to serve any moralistic or ex-cathedra notions of what is right, or what "property" truly is. Policy debates of the kinds of ideas that should be eligible for protection, or on the appropriate strength of any rights created, should focus on the public's interest in these rights and how that public interest may be served by modifying them.

The general argument advanced for the public's interest in intellectual property rights is that inventors would not otherwise undertake their inventive activity and that, since the public must somehow benefit from this inventive activity, it should be so encouraged. Just how is the public presumed to benefit? The primary assumption, albeit implicit, is that the property rights are not perfect - the owners will not be able to appropriate all the benefits from the ideas. One reason for this is that the price charged for the use of the idea will be less than the value of the benefit, for at least some of the consuming public (else the owner may not benefit at all.) Also, the rights will expire in time and everyone will use the idea freely. Furthermore, the disclosure required to obtain the rights will stimulate others to further invent. In these and other ways, the public will benefit from inventive ideas drawn forth by incentives resultant from property rights, even as the rights assure that the owner may also benefit, perhaps even substantially.

Innovative ideas with applications in agriculture have become more easy to come by due to the advance of scientific knowledge, particularly in the area of biotechnology. They have also become more valuable to their creators over the past twenty years because of the new possibilities for intellectual property rights bestowed through new breeders rights and the extension of utility patents to living things. Policy issues related to these intellectual property rights are now of two kinds: fine tuning adjustments of the rights in the U.S. and some industrial countries, and whether the rights should or will exist at all in some other countries. Public debate on these issues must focus on which public is to be served by any changes and how they might benefit, rather than appeals to notions of justice for the innovators or justice for the users (whether those parties be individuals or nations.) This paper sets forth some more explicit notions of private and public benefits that will be useful in considering the desirable strength of intellectual property rights and examines some numbers related to the effectiveness of plant variety protection rights in the U. S.

Intellectual Property Rights in the Public Interest

It was asserted above that intellectual property rights (IPR) have been created to serve the public interest. The public in question may well include the innovator, but consists more generally of the users of the potential knowledge. Stronger property rights permit the owner to more effectively exclude others from using the idea without the owner's permission. In this way, the owner can appropriate a larger fraction of the user benefits, and this increases the incentive for the idea to be brought forth. But as the inventor excludes users, diffusion is restricted and some of the public benefit will be lost. The tension between appropriability and diffusion constitutes the essential dilemma of IPR policy. Strong property rights help to ensure that ideas will be brought forth, but limit their benefits once produced. Weak property rights ensure wide diffusion of ideas once they are brought forth, but discourage the effort necessary to bring them.

This dilemma can be illustrated by considering a particular potential innovation - say a variety of soybeans tolerant of low soil pH. The net extra value of such a variety may be quite high for some fields, while it is worth nothing for others. In Figure 1, the potential use of the variety on particular plots of land is ranked from the highest-value ("a" dollars per acre) to the last acre on which the variety has any value at all. The total number of acres on which the variety has any advantage is "dm". The total economic benefit to society from this innovation is the sum of the extra values on each acre, which is the equal to the area under the curve. If perfectly enforceable property rights are available for this variety, the owner will charge some royalty rate equivalent to p*/acre for anyone who uses the variety[1], and the use of the variety will be restricted to those acres for which the extra value exceeds this royalty rate. In Figure 1, d* represents the number of acres on which it will be used, and the area under the curve to the right of d* represents potential beneficial applications of the innovation that are not realized because the property right owner has restricted its diffusion to those acres where net extra value is at least large enough to pay the royalty.

Figure 1 offers further insights into the distribution of benefits from the new innovation. The amount that the property right owner will receive in royalties is the amount of the royalty, p*, times the number of applications, d*, or the area represented by the rectangle Op*bd*. Producers (and/or the ultimate consumers) will keep the remaining total benefits, which are equivalent to the triangle p*ab above the royalty rectangle just described[2]. It is clear that the development of this new variety is in the public interest if these total benefits (royalties plus producer/consumer benefits) exceed R&D costs. To consider how many of such innovation efforts should be undertaken, we must return to the full range of potential innovations that could conceivably be developed in a given time span.

Let's say there are M such potential innovation projects, of which acid-tolerant soybeans are but one. For each, there is a potential social benefit corresponding to the area Oadm in Figure 1. If we rank these M potential innovations from highest to lowest social benefit, we have a schedule such as line AM of Figure 2. If there were no R&D costs, it would be desirable for the society to develop all these innovations. But if the R&D costs rise as more and more innovations are undertaken, then there will be a point at which the R&D cost of an additional project exceeds the social benefits, and this "social optimum" point is illustrated at point B of Figure 2.

Unfortunately, even intellectual property rights will not permit the society to reach such a conceptual bliss point. With property rights, some of the social benefits from each innovation are lost because of restricted diffusion, so the social benefits under property rights are less, represented by line EM. Furthermore, the entrepreneurs who undertake research and development are guided not by the return of social benefits over costs, but by the return of royalties over cost. The royalties earned from each of these potential innovations are, as we have seen, some fraction of the diffusion-restricted social benefits which we indicate here with line DM. Entrepreneurs will then undertake to discover that fraction of innovations whose royalties are expected to be at least equal to the cost of discovery. In Figure 2, that number of discoveries is indicated by Ka, compared to the Ks discoveries that would be optimal for the society. If the social benefits from the remaining discoveries are to be realized, some other incentives or public funding will have to be provided.

This conceptual analysis has assumed that the intellectual property rights are essentially perfect. It is instructive to consider weaker property rights, such that some consumers will be able to use the innovation without paying the royalty. In terms of Figure 1, this means that royalty rate p^* will result in fewer than d^* royalty payments, and more than d^* uses. The owner of the rights will also have an incentive to charge a lower price than p^*, thus increasing the diffusion of the innovation even further, though the royalty receipts would be smaller than under perfect property rights. The effect of weakened property rights on the rate of innovation can be seen in Figure 2, where the extra diffusion of each innovation increases marginal social benefits (line EM rotates upward towards line AM), and reduces royalty benefits (line EM rotates downward toward OM). The result is that fewer innovations will be undertaken by entrepreneurs, but the net social benefits will be larger for each of those innovations that are undertaken. From the point of view of the whole society, then, weaker property rights may result in more benefits from innovation, even though the rate of innovation is reduced even further from the social optimum level Ks.

The strength of intellectual property rights is determined by a number of factors, some of which are very well known to the plant breeding and biotechnology community. Among these are:

(a) The number of years the property rights are valid. The longer the patent life, the more of the future gains can be appropriated via royalties.

(b) The breadth of protection that is granted. The broader is the domain of "innovation space" that is protected, the greater is the fraction of the returns that can be appropriated through royalties.

(c) The severity of required enabling disclosures. The more information is required to be disclosed, the more information is available for competitors to use in "inventing around" the innovation to create a substitute.

(d) The kinds of uses that are prohibited without the owners' consent. Research uses, and sales by farmers are kinds of uses that are not prohibited under PVPA, but are prohibited under utility patents, thus weakening the former relative to the latter.

(e) The cost of enforcement mechanisms. The higher the costs to monitor the use of the innovation by others, and to enforce the right to prohibit others from using it, the weaker are the property rights.

The analysis above provides a sketch of the rationale for the public interest in intellectual property rights and for public policy related to the appropriate strength of those property rights. We now turn to some data related to the issue of the effectiveness of property rights in increasing the rate of innovation in the case of plant variety protection in the U. S.

Plant Variety Protection Rights and Plant Breeding Efforts

The U.S. Plant Variety Protection Act of 1970 established a new set of property rights. While only some fine-tuning questions are at issue in the U.S. (such as the farmer's exemption), the desirability of such rights is at issue in other countries. It is therefore useful to review the public interest at stake in PVP from the perspective of the above analysis. One basic question is the incentive effects of the Act - did it have a significant impact on the level of R&D in plant breeding. A second question is whether the strength of those rights has led to an appropriate balance between appropriability and diffusion. We are much closer to an answer to the first question than to the second.

Incentive Effects of PVP

Prior to passage of the Act in 1970, private plant breeding efforts were virtually nil for soybeans and the cereal crops. From just two such programs spending about $10,000 per year in 1960, by 1979 there were 30 programs spending over eight million dollars (Table 1). Increases in private breeding efforts also occurred in other crops during this period, and those increases are probably due in part to PVP. However, those increases are less dramatic, probably because de facto property rights already existed in hybrids (where secrecy allows some appropriation of benefits from breeding) and in vegetables (where nation-wide brand marketing of products allows some appropriation of benefits from breeding). The trend in cumulative number of varieties available (Figure 3) is a measure of the fruits of this increase in breeding efforts.

In order to determine the effects of PVP on plant breeding efforts, it is necessary to determine what other factors are affecting these efforts, so the property rights effects can be isolated (see Foster and Perrin for such an effort.) For example, even with perfect property rights in varieties, one would expect breeding efforts to be smaller for the less important crops, simply because there is less of a potential market for the seeds, and thus a smaller potential payoff for a good variety. In Figure 4, we can see that on the basis of the number of varieties certified per million dollars of crop value, there has been a great deal more crop breeding effort in soybeans and cereals than in corn and sorghum. This again suggests that PVP property rights had a great impact on breeding of crops for which appropriation of benefits was not otherwise available.

Finally, since the breeding of soybeans seems to have been most dramatically affected by PVP, it is useful to examine those trends more carefully. Figure 5 combines data on the number of private firm breeding programs and the number of PhD's employed in those programs. Both sets of numbers start from near zero and have continued to increase right up through 1989. It seems quite clear from the above kinds of data that PVP has indeed had a significant incentive effect for private-sector R&D efforts.

The Strength of PVP Rights

A key variable related to appropriability is the royalty price that owners are able to charge. In the case of PVP, this can be interpreted as the premium which owners are able to charge for protected seed, relative to non-protected varieties of seed with the same quality characteristics (germination, purity, etc.) This is the source of funds to repay R&D costs.

Little direct data on the level of these royalties has as yet been collected. An early study by Marion and Butler indicated that in the early days of PVP the premium may have been in the vicinity of 20%. Indirect evidence is that firms' expectations were that the premium would be sufficient to offset the R&D costs, else we would not have seen such a dramatic increase in private R&D activity. As the industry has gained more experience with both research costs and price premiums, this perceptions may change, but it is not evident yet.

The distribution of benefits between certificate owners and producers will be determined in part by the size of the yield superiority relative to the premium charged. Some preliminary evidence and analysis (Perrin, et al.) failed to establish that average variety test yields had been increased much by the influx of protected varieties, at least as of 1980. That analysis is being updated, with more careful attention being given to locally superior varieties, and the yields of commercially successful varieties as opposed to simply all of the varieties being tested.

Also relevant is the diffusion of protected varieties. Recall that a potential limitation of intellectual property rights in serving the public interest is that the royalties limit diffusion and therefore total benefits from the innovations. Huffman reports that about 70% of the soybean crop each year is planted to purchased seed, a large majority of which is private variety seed. Given that much of the remaining crop is probably planted to bin run seed of those same varieties, this suggests that effective royalties are low enough to insure wide diffusion.

A final indirect measure of the adequacy of the strength of rights would be measured productivity gains in production of the crop. If property rights are too strong, measures of productivity gain in the farming sector would show little improvement from private R&D, since most or all of the benefits of the new technology would be paid out by the farm sector to the seed sector in the form of royalties. However, Huffman and Evenson's analysis indicates that private sector R&D has a very large impact on crop productivity - even larger that public sector R&D. While theirs is evidence over a long period and for all crops, it suggests that there has been no general problem of excess strength of PVP rights.

Conclusions

Property rights in crop varieties and other innovations embodied in living materials serve a legitimate public interest, not just the interest of the innovators who expect to obtain the property rights. No property rights can be perfect and costlessly enforced, and in fact the public interest is served best by property rights whose strength is somewhat diluted so that wider diffusion of the innovations occur. As a practical matter, however, it is no simple matter to determine just how strong the property rights should be. On the other hand, it is important that the defense of intellectual property rights be made on the basis of the public interest, rather than on the interests of innovators alone (a "special" interest group), for it is the public who will determine whether such rights shall exist and what their strength shall be.

The property rights created by PVP seem to have been of a reasonable strength to serve the public interest. Those rights have been sufficiently strong to have greatly stimulated private R&D efforts, and the royalties charged do not appear to have limited diffusion much, although data on this are very sketchy. Probably sufficient time has passed that the social payoff from these efforts can now begin to be assessed, although this has not yet been done.

As the strength of PVP rights as well as patent rights continue to be defined through administrative, judicial and legislative action, it will be important to seek a balance in providing incentives for private R&D efforts while ensuring wide commercial diffusion of the varieties and other innovations.

Table 1

Crop Breeding Research Efforts Reported by 59 Firms[a]

	1960		1970		1979	
Crop	Programs	Expend. ($1000)	Programs	Expend. ($1000)	Programs	Expend. ($1000)
Hybrid Corn	14	1,873	18	4,913	32	19,745
Hybrid Sorghum	6	448	12	1,202	18	2,847
Soybeans	1	2	8	270	21	4,296
Cereals	1	8	6	1,083	9	4,328
Forage and turf grass	4	256	9	1,077	16	3,049
Vegetables	7	977	12	2,522	16	7,517
Other crops	1	8	6	226	11	878

[a]From Perrin, Hunnings and Ihnen

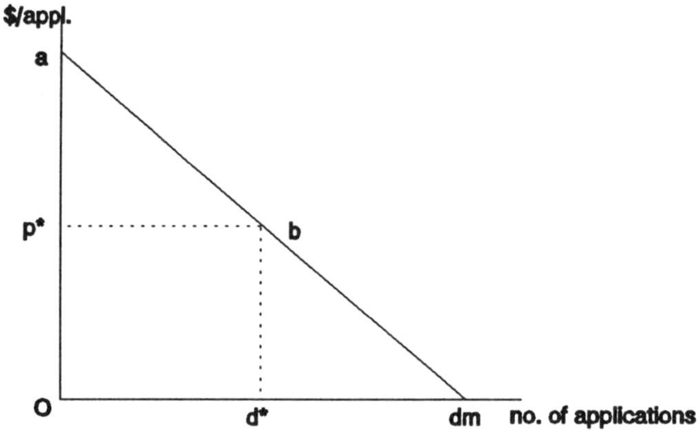

Figure 1. The derived demand for diffusion of an innovation

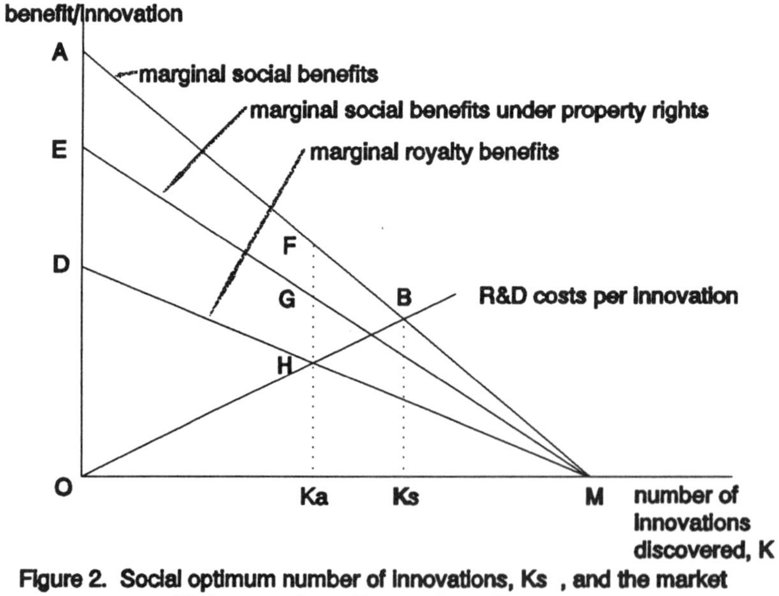

Figure 2. Social optimum number of innovations, Ks , and the market equilibrium number of innovations, Ka .

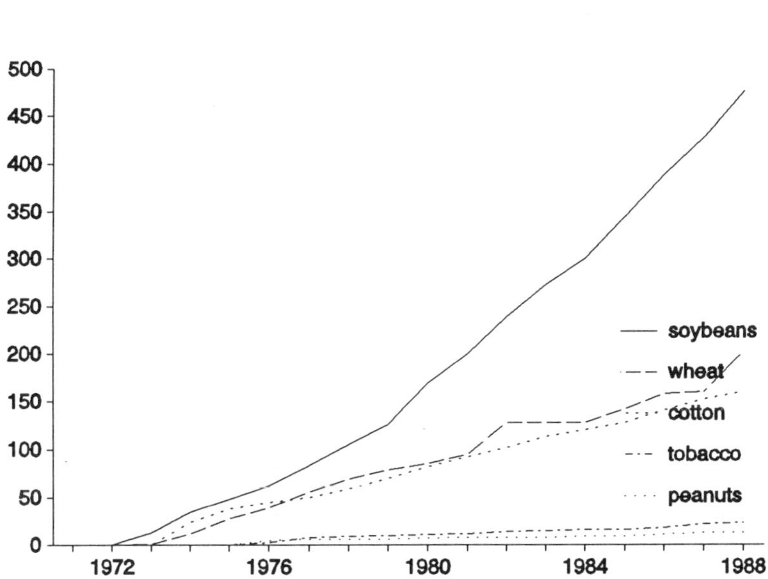

Fig 3 Cumulative PVP certificates granted for selected crops.

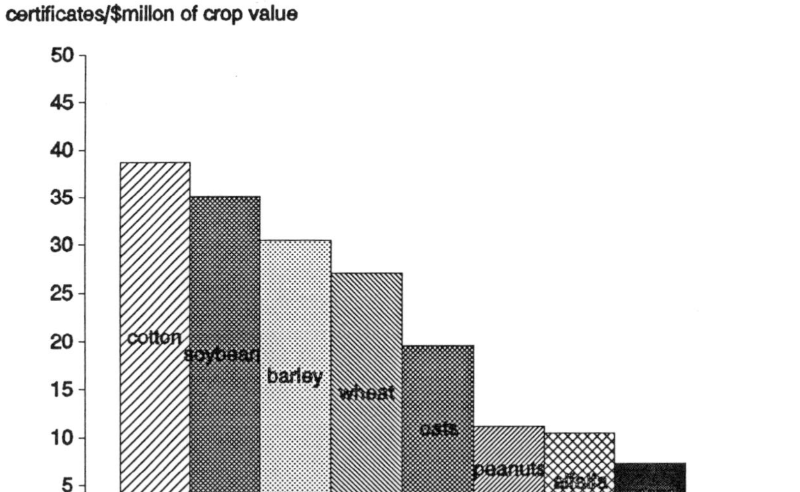

Figure 4. PVP certificates issued (by 1990) per million dollars of crop value.

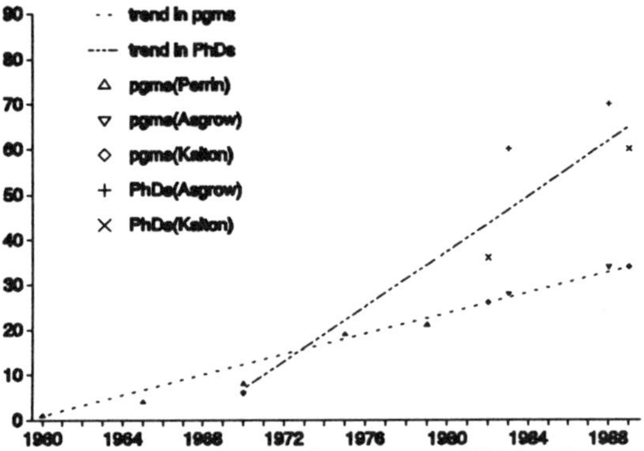

Fig 5 Number of soybean breeding progams and PhD breeders.

Endnotes

1. For our purposes here, a royalty rate is the difference between the price charged for a bushel of seed and all non-R&D costs related to the production and distribution of the seed. It is a profit over and above normal that constitutes the return to the R&D efforts.

2. The gains represented by triangle p*ab provide the initial incentive for producers to adopt the variety. Some or all of these gains may be transferred to the ultimate consumers as product prices begin to fall due to the reduced production costs made possible by this and similar innovations. The royalty returns to the property owner will persist until another new variety comes along that reduces (or eliminates) the productivity advantage of the variety being considered here.

References

Asgrow Seed Co. 1989. *A Chronicle of Plant Variety Protection - 1989 Update* Asgrow Seed Co, Kalamazoo, MI.

Foster, William E. and Richard K. Perrin. 1991. "Economic Incentives and Plant Breeding Research," *DARE: 91-05*, Department of Agricultural and Resource Economics, North Carolina State University, Raleigh, NC.

Huffman, Wallace E. 1992. "Economics of Intellectual Property Rights in Plant Materials," unpublished manuscript, Iowa State University, Ames, IA.

Huffman, Wallace E. and Robert E. Evenson. 1993. *Science for Agriculture*, Iowa State University Press, Ames, IA.

Kalton, R. R.; P. A. Richardson, and N. M. Frey. 1989. "Inputs in Private Sector Plant Breeding and Biotechnology Research Programs in the United States," *Diversity* 5(4):22-25.

Perrin, R. K.; K. A. Hunnings and L. A. Ihnen. 1983. "Some Effects of the U.S. Plant Variety Protection Act of 1970," *Economics Research Report No. 46*, Department of Economics and Business, North Carolina State University, Raleigh, NC.

Chapter 14

Modifications in Intellectual Property Rights Law and Effects on Agricultural Research

W. Lesser
Cornell University

During the 1980's, major changes were made in intellectual property law in the United States. Identifying those of particular relevance to agriculture, the fundamental extension of patent law was the Supreme Court's decision in Chakrabarty that, "anything under the sun which is made by man" is patentable subject matter.[1] The "anything" in that case was of course a living organism, a microorganism in that instance. Subsequent internal Patent and Trademark Office interpretations of Chakrabarty lead to the patentability of seeds[2] and animals[3]. Also notable, though of a different order of magnitude was the 1980 amendment to the Plant Variety Protection Act[4], the US legislation establishing Plant Breeders' Rights. Among other changes, those amendments made it possible for the United States to join UPOV, the international Plant Breeders' Rights convention.[5] At this point it can be claimed that the US has the broadest range of allowable intellectual property protection for living organisms in the world. What do the 1990's hold in store?

This decade is, first, likely to see the geographic expansion of protection worldwide. As of 1988, for which we have a good accounting, 52 countries expressly excluded patents for plants and animals[6] while about a billion of the world's people live in countries lacking effective patent legislation. The United States, joined by other developed countries, has interpreted an absence or restriction of intellectual property rights (IPR) protection as a trade barrier under

GATT - referred to as trade related aspects of intellectual property or TRIPs.[7] While the fate of TRIPs in any final GATT agreement is unclear, the US through the Trade Representatives Office has unilaterally pressed for enhanced IPR protection. That campaign is credited with success in Mexico, Indonesia, PRC and elsewhere; India may follow soon (Biotechnology and Development Monitor 1991). On what basis have these efforts been established, and what effects might they be expected to have for agricultural research in the US and worldwide? That is the first issue to be explored in this paper.

A second attribute of IPR in the 1990's might be called fine tuning of the laws. Minor changes might be made to adapt to unanticipated situations or to enhance worldwide harmonization. One recent example of this kind of change is the 1991 UPOV revision which introduced "dependency" into PBR. In brief, under this revision, if variety B is derived from A, and C from B, then both B and C are "essentially" derived from A and owing royalties to its owner. What affects will dependency have on plant breeding research in the US? In exploring that question, the second issue developed here is the formulation of a recommendation as to whether the US should accede to the revised variant. For both of these issues, it will soon become apparent, there is much that we cannot answer specifically at this point. Thus the major outcome of the current effort is a further delineation of research needs.

Geographic Expansion

Projection of the impacts of enhanced IPR protection in developing countries on agricultural R&D is based first on what is known about the impacts of such protection on R&D funding. That exploration has two components, the theoretical and the empirical. Theoretical components are treated first. Subsequently it is essential for projecting private investments to understand the strategic decisions of firms, and US firms in particular, in regards to what is spent and where. That is treated in the section Strategic Behavior. Finally, in the section titled Conclusions and Research Needs, the projections are made and research needs are identified.

Theoretical Issues

The US patent law was passed, in words attributed to Thomas Jefferson, to "promote the progress of science and useful arts, by securing for limited times to authors and inventors the exclusive rights to their respective writings and discoveries"[8]. This constitutes the economic justification for IPR in contradistinction to the more philosophical inalienable rights ("natural law") to personal creations, that "an idea belongs to its creator because the idea is a manifestation of the creator's personality or self" (Hughes 1988 P. 330). This paper considers only the more utilitarian economic incentives in line with the general interpretation of Article 1 to foster social goals (e.g., Anderfelt 1971, p. 13).

The fundamental concept of IPR, as best expressed by Machlup (1958), is the provision of an incentive through the opportunity to recover research expenditures by allowing a limited, temporary monopoly to the creator. Such a monopoly protects the creator from direct copying while necessitating that any financial rewards come from the market. The market determines the value of inventions, not the patent office. To the degree that it has been estimated, only some 5-15 percent of patents are used (practiced) to any extent (Nogués 1989; Taylor and Silberston 1973). When referring to the monopoly rights, the adjective "limited" applies to the allowed scope of the patent or certificate of PBRs. "Scope" is a legal term referring to the extent ("closeness") of close copying which would be considered an infringement of the patent (Crespi 1988, Chapter 4).[9] Obviously meaningful protection must apply to more than exact copies,[10] but the establishment of the appropriate scope is a complex matter determined in part by the patent examiner and, when necessary, by the courts.

"Temporary" refers to the duration of protection, typically about 20 years. Society extracts one other self-limitation from the applicant, the requirement that the invention be revealed, described so that anyone knowledgeable in the art can recreate (practice) it (35 U.S.C. Sec 112). This disclosure is typically satisfied by a written description but when that is not technically feasible, as is often the case with living organisms, a deposit of the protected material may be added[11] (see eg., Straus and Moufang 1990).

A more succinct way of saying all of this is that patents and other forms of IPR allow only negative rights; the right to exclude others from using your creation. IPR does not give anything, not even the right to use one's own patented invention. That invention may require regulatory approval for use, as with pharmaceuticals or the intentional environmental release of many lifeforms, or its use may infringe on a previously-existing patent. Certainly the issuance of a patent is not some form of governmental sanctioning of the invention as some commentators seem to imply.[12]

Clearly there is an effort by the formulators of the IPR laws to strike a close balance between public benefit - the bringing forth of private research efforts, and public cost - the monopoly rights granted. What does economic theory have to say about that balance? In terms of the overall system, little beyond the descriptive. However the laws contain a myriad of specifics which are more conducive to economic analysis.

Economists have subdivided the invention process into three stages:

* invention,
* innovation, and
* diffusion (commercialization).

Studies have indicated that the basic creative process, the foundation of many significant inventions, is motivated more by the creative drive than by monetary incentives. Patents are more important in the long, tedious and expensive process of making the invention commercially acceptable (Jewkes, Sawers and Stillerman 1969).

Economists have looked particularly at the duration of protection and the ramifications of the "winner take all" incentives established by patents and like laws. Considering duration, the analysis has shown, not surprisingly, that a fixed term is not optimal. The ideal term depends on the demand elasticity and significance of the invention in cost reduction (review in Primo Braga 1990). As a practical matter, this result is not very useful because we already know from patent renewal data, the escalating periodic fee to maintain a patent, that few patents are commercially viable for the full 17 plus years (Schankerman 1991). Thus the statutory life of patents is only rarely of practical importance so that, overall, the social cost of extended protection is small. Moreover, it is not administratively feasible to grant different durations of protection for different types of products.[13]

The race to be first is perhaps more conceptually interesting. At issue is the lack of reward to be second in the race. Does this cause more and wasteful investment in a race, or are firms conservative in not gambling to be first, holding down total investments? Alternatively, the ensuant competition may compel firms to work faster and take alternative approaches, enhancing the likelihood and speed of finding a useful result. There has been no general satisfactory resolution to this question because the analysis requires that strong assumptions be made about behavior and competition (Reinganum 1982; Harris and Vickers 1987; Loury 1979). But perhaps the matter is mooted by Dasgupta's (1986) observation that technological competition is a continuous process from which firms are absent at their peril. This is in contrast to the discrete decisions implicit in the issue of patent races.

Overall, in Primo Braga's (1990, p. 32) words, "Economic theory has raised more questions about welfare implications of intellectual property than it has answered. The theory of intellectual property protection is fragmented and provides no robust answer to the question of the appropriate or optimal level of protection under various sets of real-world circumstances." As limited and flawed as IPR clearly are, no one has yet identified a clear improvement (Jewkes, Sauers and Stillerman 1969; Benko 1987; Wright 1983).

Empirical Results

With the theory providing little guidance on the subject of the incentive effects of IPR, the issue becomes an empirical matter. Here most of the evidence applies to Plant Breeders' Rights, the patent-like protection for traditionally bred open pollinated (US only) plants. The apparent reason for favoring PBR is its recentness, dating as it does only to 1970, whereas the US patent law was first adopted in 1790. The more recent legislation allows for "before and after" analysis not possible when the before was ten score years ago. Plant breeding for its part provides a concise sphere of activities with an easily identified product for evaluation. However, even if activities are temporally correlated, it does not necessarily demonstrate causality.

This is not to say that no estimates of the incentive effects of patents have been attempted. Schankerman (1991) used 1970 patent "cohorts" by major sector (pharmaceutical, chemical, mechanical and electrical) to estimate the private value of patents worldwide. He found the median value by sector to range from \$1,600 to \$3,100 in 1980 US dollars. There is an indication in these figures of the "winner takes all" component, for the top one percent of patents accounted for 15 to 25 percent of the median value (depending on sector). More pertinent to the incentive issue, Schankerman found that patents generate about a quarter of the private returns to inventive activity, not inconsequential, but not dominant either. These figures confirm statements of many executives that patent protection in general is not a critical factor in making R&D decisions (Scherer 1980, p. 446; Nogués 1990, pp. 11-14).

The overall moderate significance of patents may, however, not be indicative of agriculture and especially the "new biotechnology" where for self-reproducible lifeforms legal protection is about the only protection there is. And nothing is as easily copied as non-hybrid seed where each plant has the means of replicating itself many times over. In contrast, for many other technologies portions of the know-how are kept as trade secrets so that the invention in its most efficient form cannot be readily copied.[14] Thus, experiences under PBR are potentially very revealing of the potential incentive effect of IPR in agriculture.

Prior to examining the evidence it is helpful to understand better the differences between patents and PBR. Here the US law will be examined as the empirical evidence is from there. PBR has requirements of distinctiveness (clearly distinguishable), uniformity and stability (summarized as DUS) to parallel the novelty, utility and nonobviousness requirements for patents. With PBR distinctiveness is the key requirement, as it defines to a large extent the scope of protection. Uniformity and stability are technical requirements to assure that the variety has been bred for a sufficient number of generations to breed true to type. The distinctiveness requirement in the US is interpreted very narrowly so that virtually any difference, even if of no practical value at all, is sufficient to receive protection. This interpretation severely limits protection from close copying to the point that, in my interpretation, the US law really protects the variety name, not the germplasm itself (Lesser 1986).

PBR has another clause limiting the scope of protection. This is the "farmer exemption" (7 USC Sec. 2544) which permits farmers the right to retain seed for replanting and to sell limited quantities as a secondary activity.[15] Farmers then become direct competitors to the seed companies, limiting the value of protection. I personally believe this is not such a major factor, as farmers will buy new seed anyway every third year or so because of genetic drift and improvements in the intervening years. Moreover, companies can price a three year input higher than a single year input (Lesser 1986). Nonetheless, in an empirical study Knudson and Hansen (1991) found that farmers could increase yields and profits by using purchased rather than "bin run" seed. However because farmers do use bin run seed for about a third of their needs, the farmers privilege does reduce the value of PBR.

Against this backdrop of limited protection, how has the private sector responded? Overall, quite substantially, as measured from several perspectives. Butler and Marion (1985, Chap. 3) conducted a survey of major private seed companies' investments in nonhybrid breeding before and after the passage of PBR in the US and found for the 14 largest firms a quadrupling in constant dollars of R&D expenditures from 1960-80. Over the same period there was a marked shift from corn breeding, which as a hybrid cannot be recreated through saved seed, to open pollinated varieties, especially soybeans. Numerous smaller firms established open pollinated breeding programs and employed breeders over the same period. Most of the major investments were made prior to the 1970 passage of the PVPA, so it must be inferred that this was done in anticipation of passage. That is, causality cannot be proven. Their findings are summarized as follows (Butler and Marion 1985, p. 30):

> *The sharp increases in R&D expenditures that occurred during 1967-70 were largely concentrated on research facilities and non-personnel costs. This was followed during 1970-75 by a significant increase in the number of plant breeders per firm. Since 1975, plant breeding R&D activity has experienced more modest increases.*

It should be emphasized that the effort has been uneven across crops with soybeans receiving the greatest attention, followed by wheat, while most crops are unaffected. Perrin, Hunnings and Ihnen (1983) have shown that the level of private investment across crops can be explained by such factors as the yield increase potential and the multiplication rate, as well as the value of the crop. Prior to the passage of the PVPA essentially all non-hybrid breeding was done by the public sector. Subsequently, it is important to note that there was no obvious reduction in the public effort in this activity, although budgetary limitations are clearly having impacts across the spectrum of public sector activities.

Brim (1987) and Foster and Perrin (1991) have provided further data on increases in breeding programs and breeder numbers following the passage of the PVPA. Private programs have increased from one to 34 over the period 1960-88 while Ph.D.'s employed by private firms grew from six in 1970 to seventy in 1988.

Several studies have attempted to draw inferences from the large increases in numbers of certificates of plant variety protection which have been granted. However as Stallman and Schmid (1987) point out, the mere existence of certificates connotes little, because many will represent modest and practically insignificant differences. The breeding of so-called cosmetic differences has been an issue since 1980. Perrin, Hunnings and Ihnen (1983) have given some insights into this issue by showing that private varieties are somewhat more productive than public varieties, although the statistical evidence is limited. It can also be noted that for some crops, including soybeans, private varieties now dominate acres planted. If it can be assumed that farmers are good appraisers of varietal differences, which seems likely in a competitive sector with measurable product attributes, than this too suggests that private varieties are productive.

Overall the evidence supports the theoretical projection that the incentives provided by IPR do indeed stimulate private R&D investments. There remains much that is not known, including the level of incentive required, differences across sectors, and interaction with other protection mechanisms including secrecy. Yet when secrecy is not possible, as with self-reproducible organisms, there is a clear private response to IPR. The net social return to this protection is a far more difficult question to answer, but that is not the focus of this paper.

Strategic Behavior

If the above is convincing in indicating the existence of positive private incentive effects of IPR at least for certain agricultural products, then the ancillary questions for the US agricultural sector are: 1) does greater protection further encourage private research by expanding the geographic markets? and 2) where geographically are those monies spent? That is, what are the strategic decisions made by firms in determining where and how research funds are to be used? Clearly, it is significant for the US and for our university system that the work be done domestically.

This, regrettably, is an area about which we know little. Firms must balance multiple factors, including:

* The availability of trained researchers,
* The product life cycle (Bozeman and Link 1983), and
* Local adaptation requirements (Evenson 1988).

For agriculture several of these factors create divergent forces. Typically agricultural inputs, especially seeds, require very local adaptation, necessitating that the work be done within a wide geographic scope. On the other hand, the life cycle for most self-reproducible agricultural inputs is brief, suggesting that firms will attempt to license them widely. In the words of a textbook explanation (Cundiff and Hilger 1984, p. 290):

> *If the life cycle of a technology is expected to be short, it is probably best to use quick methods such as licensing to introduce the product, process or technology to the widest possible world market at the earliest possible time.*
>
> *Since patents are an assist to licensing this suggests that a geographical extension of protection will enhance domestic R&D.*

Some overseas markets are facilitated by IPR. Speaking for Canada as an importer, most likely from the US, Young (1989) described the situation with plant varieties as follows:

> *Some private varieties should be available for use in Canada even though they may be created elsewhere as part of the larger plant breeding program, and a consequence of no breeders' rights*

legislation is a restriction on the availability of such private varieties.

The same situation applied to Argentina which has purchased varieties from the US following the passage of its own PBR legislation (Gutiérrez 1991). Hence geographically expanded protection can be expected to enhance the total market, but probably not greatly for many agricultural products due to local adaptation requirements.

The subsequent issue is where geographically that research will be conducted. Overall the location of R&D is likely to change in future years as the number of well trained researchers (many in US universities) provide less expensive options to working in the US. These choices will not be available for complex products where facilities in developing countries are inadequate, but that applies to a limited number of technologies. Indeed, IPR may make a shift in research activity more possible as research operations depend heavily on maintaining secrecy, or "trade secrets" to use the formal term. Or there may be a more direct link, as with the apparent agreement between Canada and the major pharmaceutical companies to expend research funds there in exchange for the allowance of patents for those products (Spurgeon 1992). Assuming no major increase in overall funding, this means expenditures elsewhere, including in the US, will be reduced.

Conclusions and Research Needs

The available evidence, as limited as it is, suggests that IPRs do what the theory predicts, they foster private R&D. The incentive effect, to the degree it can be inferred, appears not to be great overall. Living self reproducible organisms, for which there is little protection beyond the legal, are an exception; protection is critical to private firm involvement. Because agricultural research includes a large portion of such products, the importance of IPR is consequently greater than the average. And we have fairly strong evidence that appropriate protection increases private domestic R&D.

The US presently has, by world standards, broad and strong protection so that little will change domestically to enhance incentives and investments. The attention is now on the enhancement of IPR protection elsewhere, notably in developing countries, which often have weak or nonexistent protection. When particular products are identified for exclusion, they fall heavily in the realm of agriculture, such as a ban on "plant or animal varieties". Pharmaceuticals, too, are frequently excluded. Some advances have been made recently in response to pressure from the US in expanding protection in such countries as Mexico, China and Indonesia.

The indications are that these legislative changes will lead to some, but not major, increases in agricultural research in the US. Many agricultural applications require local adaptation restricting the direct market. What added research is done may be increasingly concentrated in the target markets, especially in developing countries where costs are lower. The enhancement of IPR can hasten that movement, but the key factor is the training of foreign researchers in major universities, especially in the US.

To understand more fully the balance of these factors and their implications on US agricultural research, more needs to be known about the incentive effects of IPR. However those issues are related principally to the social optimization of IPR, certainly an important issue but not the focus here. Where our knowledge is really lacking is in understanding where research moneys are spent and the affect of IPR on those decisions. The US through the Trade Representatives' Office has moved ahead vigorously on strengthening IPR in developing countries. This may help multinational firms overall, but its affect on research within the US while not well understood, is likely to be limited.

Fine Tuning

The statement was made above that IPR in the United States is relatively broad and strong by world standards and that a major target of US and other developed country policy in the subject area is the enhancement of IPR protection in developing countries. At the same time, it was concluded that it has not been possible to determine whether current protection is optimal. None of this, however, prevents ongoing efforts to modify protection in specific aspects. The aspect under consideration here is a recently proposed modification of UPOV, the international convention for PBR. This convention, the International Convention for the Protection of New Varieties of Plants, goes into effect when signatory nations adopt similar legislation into national law. In the US, the applicable law is the Plant Variety Protection Act of 1970. The US joined UPOV in 1980 following some minor harmonizing amendments.

When UPOV was first drafted in 1961, it appeared to function as a "separate but equal" statute for plants which paralleled patent protection for mechanical, chemical and electrical inventions. This apparent specialization of protection, or as it is commonly known, double protection, is evident in Article 2(1) which reads,

> *"Each member State of the Union may recognize the right of the breeder provided for in this Convention by the grant either of a special title of protection or a patent. Nevertheless, a member ... may provide only one of them ..."*

It should be noted that the US qualified under Article 37 of the 1978 text thus allowing it to offer two forms of protection, patents and PBR. But that is a detail; the significant point is that the Convention was first passed in an era when plants and plant-based research were clearly distinguishable from other forms of research.

That era passed quickly with the advent of biotechnology and subsequently the terminology "plant and animal varieties and essentially biological processes for the production of plants and animals" (EPC Article 53 (b)) have caused great definitional problems (see Bent, Schwaab, Conlin and Jeffery 1987, Chapter 4; Commission of the European Community 1988). At the same time, the existence of the farmers' and research privileges in PBR has meant that protection is not as strong as for patents. Using an existing protected variety as the basis for developing a new one, as is permitted by the research exemption, means that near direct copying is possible. This is true in my estimation even outside of the US where the inventive step requirement is greater (Lesser 1987a). The proposed 1991 amendments to UPOV seem to me to be an effort to restore the attempted equivalence between PBR and patents.

The amendments, which do not become law until they are ratified in a member country, apply to four major areas (UPOV 1991):

* limited extension of rights to harvested material and products thereof (Article 14(2) and (3))

* dropping of the ban on double protection (Article 2),
* making the farmers privilege optional under national law (Article 15(2)),
* mandatory extension of protection to all genera and species within five years (Article 3(1)), and
* institution of the concept of "dependency" (Article 14(5)).

This paper considers only the ramifications of dependency as the most far-reaching charge. The limiting of the subject should not be taken as an indication that the other major amendments are insignificant or not important topics of research in their own rights.

Under the 1991 version, dependency mandates that permission will be required for commercializing protected varieties "which are essentially derived from the protected variety where the protected variety is not itself an essentially derived variety." A variety shall be considered to be essentially derived from another variety ("the initial variety") when "(1) it is predominately derived from the initial variety, or from a variety that is itself predominately derived from the initial variety, particularly through methods which have the effect of conserving the essential characteristics ..." The text goes on to identify natural selection, induced mutation, variant selection or transformation by genetic engineering, as examples of methods of conserving the essential characteristics.

To my reading, this means that if variety A is recognized as the initial variety, then breeders of varieties B derived directly from A, and C from A via B, both require permission from A's owner to commercialize their developments. Typically such permission is granted for the payment of a fee, the royalty, although this need not be the case; A's owner may wish to retain direct control. Certainly, as a result of this new article it can be anticipated that there will be major definitional squabbles over such terms as "essentially derived" and "essential characteristics". While the clarification of these definitions will be important in individual applications, the overall intent is clear, at least to my mind. It appears that PBRs are being strengthened in another attempt to equilibrate them with patent protection. Biotechnicians can do their thing at the genetic level; breeders will have their protection through PBR even if engineered genes are incorporated into varieties. In most instances, traditionally bred plants lack the nonobviousness (inventive step) needed for patent protection.

It should be recognized that the concepts of "initial variety" and essentially derived" are physical descriptors of heredity and genetic relationships and are not necessarily related to PBR. The relationship arises only if varieties in these categories are also protected under UPOV (Hunter 1992).

This new article creates powerful new incentives for the way plant breeders operate. What can be surmised about the effects of these incentives on plant breeding and on the public? Two distinct cases can be considered which will be referred to as commercial breeding and background breeding.[16]

Effects on Commercial Breeding

Commercial breeding is defined as the minor year-to-year improvements which are intended for immediate commercial use. That is, resistance may be added for a new strain of a virus in an ongoing process as the viruses themselves mutate. The annual changes are typically small (considering the number of attributes of the variety) but nonetheless important, especially cumulatively. To these annual changes are attributed about half of the long run one to two percent annual yield increases for the major crops. That growth rate, when compounded, leads to doubling every 20 or so years. From the perspective of the breeder, the improvement is continuous, with no individual making a notable single contribution. Indeed at my university, Cornell University, which has a major plant breeding program, it has been the policy for revenues from PBR to go to supporting the research program rather than to an individual. It is felt that the last breeder made only a limited contribution to the entirety of the variety and is undeserving of the returns for all of it (Lesser 1987b).

Dependency will change these incentives by providing a major incentive to be first, to develop the initial variety. To the developer of that initial variety go a disproportionate amount of the returns, for to him/her go some share of the value of all dependent varieties. Breeders of the dependent varieties (B and C in our example) for their part are no better off, and possibly worse off, than presently. While their variety is marketable, they must pay a royalty to the owner of the initial variety and when it becomes outmoded,[17] as when B is improved upon by variety C, they will receive no royalties, as is the case presently. Incentives, then, are shifted heavily to the developer of the initial variety and away from successive breeders.

In cases in which the initial breeder added nothing more than the successive ones, this system is inequitable, for one breeder earns more as a result of when sequentially the award was granted rather than the market value of the contribution itself. But IPR is about economic incentives, not equity. The incentive structure is changed in major ways. What effects might that have?

Clearly there will be a rush to be first, so that the initial investment can be expected to increase. Subsequently the none-too-great PBR incentives will decline so that a decline in derivative (from the initial variety) breeding can also be anticipated. This is clearly counter to the intent of IPR. At the same time, private firms will have the incentive to develop alternative initial varieties by returning to an unprotected variety as a base. That is, if variety A was developed from "CU1", the breeder of variety B has a direct economic incentive to return to CU1 rather than use A as the base variety. This will mean to B's breeder no payments to A's owner and the possibility of receiving royalties in the future. It will also mean a reduction in cumulative breeding practices which have been so beneficial for agriculture. To my mind this aspect of the response to the dependency clause is contrary to public and private interests.

Effects on Background Breeding

Background breeding refers to the longer term process of introducing important new traits discovered in the wild into commercial varieties. The introduction of dwarfing genes, now nearly ubiquitous, into wheat and rice is a well known example of this important practice. Presently there is interest in making corn a perennial using a distant relative recently discovered in Mexico. If commercial breeding takes a few years, background breeding requires 15 or more with no promise of eventual success.

Current PBR legislation provides essentially no economic incentive for background breeding. The first dwarfed wheat variety, for example, could be improved in some other minor, unrelated way (eg., rust resistance) and protected in its own right. The background breeder would collect royalties for a few years at best while competitors were developing their improvements. For this reason, background breeding is done principally at public expense, although with declines in public funding the amount may be ebbing. Dependency would change that.

The background breeder would (should) qualify for initial variety protection and receive royalties for an extended period. The financial incentive would be far greater for this important activity, and the amount should increase. This, I believe, is what was intended for enhancing incentives when the dependency concept was adopted.

Conclusions and Research Needs

The new dependency clause in the revised UPOV statute fundamentally impacts breeding incentives in two ways. For minor year-to-year changes in commercial varieties it shifts the benefits to the first protected, while reducing the incentives for subsequent enhancements. This creates some start up anomalies which should be considered. For example, a new variety may be dependent, stretching back 10 or more years through a series of essentially derived varieties. It will be extremely difficult for the breeder in 1993 to determine whether dependency exists and whether his/her new variety can be commercialized and at what cost. For the owner of this ageing initial variety the dependency payments will come as a windfall because the investment decisions were made before they existed. It is contrary to the concept of intellectual property rights as a form of investment incentive to allow financial rewards for something which would have existed anyway. Moreover and more importantly, dependency payments create the incentive for breeders to abandon the long term practice of cumulative improvements in favor of the development of parallel lines, each based on a firm's own initial variety.

On the other hand, dependency establishes incentives for longer term "background" breeding which are totally lacking under current PBR. The lengthy process of introducing important new traits from the wild into commercial varieties is imperiled by the lack of private firm incentives and the declining budgets for the public sector. Dependency would do much to correct that.

How should these two divergent incentives be balanced? The extent to which they exist is based on what can be called the "dependency distance," the minimum distinction from existing varieties to qualify as an initial variety (whether protected or not). This "distance" must be somewhat greater than the distinctiveness requirement for protection (UPOV Article7). A very broad dependency distance will mean substantial incentives for background breeding but also great incentives for competitors to develop their own lines based on proprietary initial varieties. The proper balance will be difficult to achieve and, in any case, will vary be genera and family. An indication of the proposed balance will come from the report on the "Draft Standard Guidelines" from the Council of UPOV. Final word, however, must be established by the case law as discussions wind their way through the legal system.

One possibility is that the "dependency distance" can be defined as greater than a generational change, something requiring effort over several years. In that case, a private breeder would have an incentive to hold back annual improvement for several years until the cumulative change was sufficient to qualify it for "initial variety" status. While this strategy would have private benefits it would slow the public's timely access to improvements. One of the theoretical benefits of the "winner takes all" aspect of IPR is that they encourage early development and release of inventions (Jewkes, Sauers and Stillerman 1969).

A second factor to consider is the implications of dependency for genetic diversity. Presently many major food crops are based on a limited number of initial lines. This means great similarity in their genetic makeup so that susceptibility to disease is potentially widespread (see National Academy of Sciences 1972). The dependency system, to the extent it interrupts the current successive selection of the best varieties for subsequent breeding, will help to broaden the genetic base of our crops. Certainly encouragement of background breeding will bring new genes into the commercial plant population. However, it should be noted that much of the concern over genetic vulnerability is based on the near ubiquity of single dwarfing genes in such major crops as wheat and rice. Additional background breeding could help in that respect, and should be considered in any debate over genetic uniformity. However, it is clear that much uniformity is driven by market forces which will be largely unchanged by the revised UPOV system.

Without, to date, a scope requirement for dependency, should the US adopt the new version or continue unchanged? To my mind, there is a major flaw in the way PBR operate with respect to protecting background breeding. PBR protection reads to the entire plant in all its attributes. This is problematic because, if what the breeder has done is to add a dwarfing gene, then that is the contribution to be recognized and rewarded. The breeder had little to do with the remainder (the great bulk) of the germplasm yet under PBR is granted some ownership rights to it. Patents are better suited to such cases as a claim would read approximately, a variety of XXX with the attribute of dwarfism. Protection is for dwarfism, potentially applicable to multiple varieties if not genera, not the entire variety. Depending on the particulars of the patent application, varieties exhibiting dwarfism would (to use UPOV terminology) be dependent on the patent and require permission from the patent holder. The use of the remaining genetic material would not be restricted. Dependency could potentially apply to multiple crops depending again on the specifics of the patent.

Patents, in this regard, are much clearer and more straightforward than initial and essentially derived varieties under UPOV- Article 14 (5). With patents it is not necessary to define such complex terms as "essential characteristics" nor to trace derivation ("essentially derived"). The partial definition of essentially derived based on the method of derivation is a further complexity because these procedures must be defined and redefined as they, too, evolve over time. The European experience with defining "essentially biological processes" from the European Patent Convention Article 53 (b) referred to above should be warning enough to avoid complex terms in intellectual property law. Certainly the revised Convention is complex because it contains two classes of protection with in the same law. Examiners (or the courts) must make a double judgment while applicants are not permitted the strategic decision of which form of protection to seek as in selecting between PBR and patents.

Complexities create uncertainties and nothing reduces the incentive to invest more than the uncertainty of protection and, especially, the unclear right to sell an invention once completed. The patent system is less complex than the new dependency scheme and for me is preferable. There is no inherent reason why the method of development affects the patentability of an invention. The patent requirements of novelty, utility and nonobviousness (inventive step) are neutral as to the means of reaching the objectives, whether it is the technically most sophisticated genetic engineering techniques or long hours and a careful eye in selecting among crosses in the field.

With this preference to retain patents and PBR in near their current forms as distinct means of protecting different kinds of inventions, I remain unconvinced of the merits of the 1991 UPOV text. Different kinds of inventions refers to those specific to a single variety and those potentially applicable to multiple varieties. These distinctions are already made in the law and require no changes in legislation (although a reassessment of nonobviousness may be advisable). UPOV in effect attempts to allow a level of IPR protection between current PBR and patents, but that fine a distinction may be too complex definitionally to be feasible.

A broader understanding of UPOV must await an evaluation of the effects of the other changes. If at that point the amended text is adopted, dependency payments to varieties protected before adoption should be prohibited. Should, in the final analysis, the US decide not to adopt the 1991 text, further consideration of changes in the Plant Variety Protection Act is needed. I recommend enhancing the distinctiveness requirement and a prohibition of farmer sales of protected seed, a right unique to the US (7 U.S.C. Sec. 4544).

Conclusions

Following a turbulent decade in the enhancement of IPR in the United States, the system is now focused on more modest changes. This paper considers two: the geographic expansion of protection, especially to developing countries, and the modified PBR convention.

The US and its developed country allies have succeeded, through trade access pressure and other means, in expanding IPR protection in several key countries. Others are sure to follow. While these changes are likely to help certain sectors, particularly pharmaceuticals and publishing, the benefit to US agricultural research is not clear. Within agriculture, living organisms require local adaptation so that much of any additional investment would be done outside the country. Other products like pesticides are relatively difficult to copy so that IPRs are not as critical. Indeed, improved IPRs may increase the opportunity to conduct agricultural research outside the US as secrecy is fundamental in research and IPRs are fundamental to secrecy. In general, agricultural research may be shifting to developing countries as lower cost options, now that well trained researchers are in place in numerous countries. Overall, enhanced IPR may be beneficial overall and beneficial to multinational agricultural firms in particular, but possibly at the expense of research within the US. This consideration should not cause us to oppose enhancements in protection but rather to plan better for their indirect consequences.

The revision to UPOV to incorporate dependency is a narrower issue to consider but nonetheless one on which a decision must be made to retain the current statute or adopt the revised one. Dependency, the stipulation that the breeder of an "essentially derived" variety seek permission for commercialization from the owner of the "initial variety", sets up divergent incentives. On the one hand, it provides a monetary incentive for long term "background" breeding, the introduction of significant new traits. This is much needed as it is completely lacking from current protection. On the other hand, dependency creates the incentive within year-to-year "commercial" breeding to breed the initial variety and not to proceed with the long term practice of small cumulative enhancements within a variety. That would be a fundamental change and on that basis alone suggests the need for extreme caution in adopting this version. However, reducing the incentive for improving only the best varieties will indirectly increase the genetic diversity of our major crops. With diversity already very limited, many observers consider our food base to be imperilled should a disease arise for which a ubiquitous gene construct has low resistance. Quantifications of either the effects of dependency on diversity or the specific risks of uniformity have, however, proven difficult so that policy makers must make a decision based on limited information. The new UPOV version contains other largely beneficial changes which need to be evaluated as well before a final decision is made. But should this version not be adopted, other approaches to strengthening PBR protection are called for and should be considered.

Endnotes

1. Diamond vs. Chakrabarty 447, US 303, 206 U.S.P.Q. 193 (1980).

2. Ex parte Hibberd 227 U.S.P.Q. 443 (P.T.O. Bd. Pat. App. & Int'f 1985).

3. Ex parte Allen 2 U.S.P.Q. @d 1425 (P.T.O. Bd. Pat. App. & Int'f 1987)

4. 7 U.S.C. sec. 2321 et seq.

5. In English, International Convention for the Protection of New Varieties of Plants.

6. World Intellectual Property Organization data summarized in Lesser 1991, Table 1. Regarding the exclusion of plants and animals from protection, it should be noted that the European Patent Convention in its Article 53 (b) contains such a prohibition although this has not prevented the issuance of such protection for plants so long as they are not in the form of a single plant variety. For developing countries the ban seems more broad based.

7. For an overview of the issues see Primo Braga 1990.

8. US Constitution, Article 1, sec. 8.

9. Technically, patent scope is determined by the claim in conjunction with the review of prior inventions, the novelty search.

10. However I have argued elsewhere (Lesser 1986) that PBR as interpreted in the US do not prevent near copying, that they protect little more than the variety name. PBR scope in other countries is greater especially as in EC when combined with commercialization requirements. See Lesser 1987(a). Patents can be informally classified into several groups, including process, product by process, new use and per se. The scope of protection varies across these types with per se allowing the greatest protection for the inventor because control is extended to new (unanticipated) uses of the invention.

11. In the case of plant patents in the US, patents for asexually propagated materials, the disclosure requirement was changed from "full, clear and concise" to "description is as complete as is reasonably possible" so that deposits are not used (35 USC Secs. 112 and 162).

12. It should be noted that in European patent law the violation of public morals ('ordre public') is grounds for rejection of an application (European Patent Convention Article 53(a)). This article was used in part as justification for the initial rejection of the initial 'Onco mouse' application there (European Patent Office 1990).

13. However in a gross sense this is sometimes done because protection for very minor inventions known as petty patents or utility models are of short duration, typically five years. For regular (or utility) patents several countries have extendable durations based on a certain requirements (Lesser 1991).

14. Many countries do not require that the patent application disclosure constitute the "best use" of the invention, only that it is a workable example. This allows the inventor to protect the most valuable form or use of the invention. US Patent Office practice, however, mandates that the best use be disclosed.

15. The revised UPOV statutes make the farmers' privilege a matter for national law, see Section III following. Presently only the US allows the sale of seed which has proven to be a difficult activity for the seed companies to detect and control. There have been calls for its repeal (American Society of Agronomy 1989).

16. I would like to acknowledge W. R. Coffman's role into bringing the key distinction between these two cases to my attention.

17. Commercial varieties of major crops have a market life of about seven to nine years and declining (Studebaker 1982).

References

American Society of Agronomy. 1989. Intellectual Property Rights Associated with Plants. Madison, WI, ASA Special Publication No. 52.

Anderfelt, U. 1971. International Patent Legislation and Developing Countries. The Hague: Martinus Nijhoff.

Benko, R. D. 1987. Protecting Intellectual Property Rights. Washington, D.C.: American Enterprise Institute.

Bent, S. A.; R. L. Schwaab; D. G. Conlin; and D. D. Jeffery. 1987. "Intellectual Property Rights in Biotechnology Worldwide." New York: Stockton Press.

Biotechnology and Development Monitor. 1991. *Plant Varieties Patentable in Mexico*. 9, 20.

Bozeman, B. and A. N. Link. 1983. Investments in Technology. New York: Praeger Publishers.

Brim, C. 1987. *Plant Breeding and Biotechnology in the Unites States of America: Changing Needs for Protection of Plant Varieties*. Pp. 117-133 in Proceedings, Symposium on the Protection of Biotechnological Inventions. Geneva: WIPO.

Butler, L. J., and B. W. Marion. 1985. "Impacts of Patent Protection in the U.S. Seed Industry and Public Plant Breeding." University of Wisconsin, NC-117 Monograph 16.

Commission of the European Community. 1988. *Proposal for a Council Directive on the Legal Protection of Biotechnological Inventions*. COM (88) 496 final - SYN159, Brussels.

Crespi, R. S. 1988. Patents: A Basic Guide to Patenting and Biotechnology. Cambridge: Cambridge University Press.

Cundiff, E. W. and M. T. Hilger. 1984. "Marketing in the International Environment." Englewood Cliffs, NJ: Prentice Hall.

Dasgupta, P. 1986. *The Theory of Technological Competition*. In J. E. Stiglitz and G. F. Mathewson, eds., New Developments and the Analysis of Market Structure. Cambridge, MA: M.I.T. Press.

European Patent Office. 1990. Board of Appeals, Case No.: T 19/90-3.3.2, Decision of 3.

Evenson, R. E. 1988. *Technological Opportunities and International Technology Transfer in Agriculture*, Chapter 7 in G. Antonelli and A. Quadrio-Curzio (Eds.), The Agro-Technological System Towards 2000. New York: Elsevier Science Publishers.

Foster, W. E. and R. Perrin. 1991. *Economic Incentives and Plant Breeding Research*. Working Paper, North Carolina State University, DARE 91-05.

Gutiérrez, M. B. 1991. *La Legislacion de derechos de obtenor en la Argentina: Una analisisde su funcionomiento y sus efectos*. In Proceedings, Industrial Property Protection of Biotechnical Inventions and Germplasm use Policies in Latin America and the Caribbean. Inter-American Institute for Cooperation on Agriculture, San José, Costa Rica.

Harris, C. and J. Vickers. 1987. "Racing with Uncertainty." *Review of Economics and Statistics*. 54, 1-22.

Hughes, J. 1988. "The Philosophy of Intellectual Property." *Georgetown Law Journal* 77, 287-366.

Hunter, R. B. 1992. "Essentially Derived and Dependency, Some Examples." American Seed Trade Association, Washington, D. C., Variety Identification Steering Subcommittee, June 8, 1992.

Jewkes, J.; D. Sauers; and R. Stillerman. 1969. The Sources of Invention. 2nd Ed. New York: Norton.

Knudson, M. K. and W. P. Hansen. 1991. *Intellectual Property Rights and the Private Seed Industry*. USDA, Economic Research Service, Agricultural Economics Rpt. No. 654.

Lesser, W. 1986. *Patenting Seeds in the United States of America: What to Expect*. Industrial Property 25, 360-367.

Lesser, W. 1987. (a) *Anticipating UK Plant Variety Patents*. European Intellectual Property Rev. 3, 81-85.

Lesser, W. 1987. (b) *Financing University Research through Patenting and Licensing: Recent Policies and Practices at Cornell University as an Example*. International Review of Industrial and Copyright Law, 18, 360-71.

Lesser, W. 1991. Equitable Patent Protection In the Developing World: Issues and Approaches. Christchurch, N.Z.: Eubios Ethics Institute.

Loury, G. 1979. Market Structure and Innovation. *Quarterly Journal of Economics*. 194, 429-36.

Machlup, F. 1958. *An Economic Review of the Patent System*. Study of the Subcommittee on Patents, Trademarks and Copyrights, Committee on the Judiciary, U.S. Senate, Study No. 15.

National Academy of Sciences 1972. Genetic Vulnerability of Major Crops. Washington, D.C.

Nogués, J. 1990. *Patents and Pharmaceutical Drugs: Understanding the Pressures on Developing Countries*. Washington, D.C.: The World Bank, WPS 502.

Nogués, J. 1989. *Notes on Patents, Distortions and Development*. World Bank (Mimeo).

Perrin, R. K.; K. A. Hunnings; and L. A. Ihnen. 1983. *Some Effects of the U.S. Plant Variety Protection Act of 1970*. North Carolina State University, Econ. Res. Rpt. 46.

Primo Braga, C. A. 1990. *Guidance from Economic Theory*. Chapter III in W.E. Siebeck (ed.), Strengthening Protection of Intellectual Property in Developing Countries. Washington, D.C.: The World Bank.

Reinganum, J. 1982. *A Dynamic Game of R&D: Patent Protection and Competitive Behavior*. Econometrica 50, 671-88.

Schankerman, M. 1991. *How Valuable is Patent Protection? Estimates by Technology Field Using Patent Renewal Data*. London School of Economics, Center for Economic Performance, Discussion Paper No. 46.

Scherer, F. M. 1980. *Industrial Market Structure and Economic Performance*. 2nd Edition. Chicago: Rand McNally.

Spurgeon, D. 1992. *Canadian Drug Firms Boost R&D Spending*. Nature 355, 666.

Stallman, J. I. and A. A. Schmid. 1987. "Property Rights in Plants: Implications for Biotechnology Research and Extension." *American Journal of Agricultural Economics*, 69, 432-37.

Straus, J. and R. Moufang. 1990. "Deposit and Release of Biological Material for the Purposes of Patent Procedure." Baden-Baden, Germany: Nomos Verlagsgesellschaft.

Studebaker, J. 1982. *Fifty Years with Breeders Rights and Plant Variety Protection*. Speech to the California Seed Trade Association, Las Vegas, Nevada.

Taylor, C. T. and Z. A. Silberston. 1973. *The Economic Impact of the Patent System*. Cambridge: Cambridge University Press.

UPOV. 1991. *International Convention for the Protection of New Varieties of Plants*. Geneva, DC/91/138, 9 March, 1991.

Wright, B. D. 1983. "The Economics of Invention Incentives: Patents, Prices and Research Contracts." *American Economic Review*, 72, 691-707.

Young, S. 1989. "Testimony on Bill C-15, An Act Respecting Plant Breeders Rights." Issue No. 5, House of Commons, Nov. 2.

Chapter 15

Research Collaboration With the Third World: Has the Free Flow of Plant Genetic Material Come to an End?

Wolfgang E. Siebeck
Washington, DC

International agricultural research draws heavily on the cooperation of the global research community. Free exchange of information and germplasm among researchers and breeders the world over has been the rule to this day, and has served research collaboration well. This is changing: while plant variety protection laws and treaties largely allow scientists to operate by the old rules, patent protection no longer does. This may not all be detrimental. Positive spin-offs should be expected.

Nonetheless, a very serious concern is the response of the Third World. In 1983, as part of the FAO "International Undertaking on Plant Genetic Resources"[1], both developing and developed countries agreed that plant genetic resources are part of the heritage of mankind, and available without restrictions to all. The Convention of Biological Diversity[2] signed in Rio de Janeiro in June 1992 by all attending industrial and developing countries with the exception of the United States recognizes sovereign rights of nations over their germplasm resources. In fact, only in exchange for industrial country funding of conservation efforts and transfer of technologies, have developing countries agreed to allow continued access to their biological resources. Hence, what used to be open to free access will no longer be. While regulations and protocols remain to be drawn up under the new Convention, if administered in an adversarial spirit, the Convention could impede the flow of germplasm in international research cooperation, to the detriment of both developing and industrial countries.

This chapter reviews research collaboration with developing countries under the traditional rules of free exchange of germplasm and research information and as it may evolve in future, under the regime of the Biodiversity Convention.

Research Collaboration under the Free Exchange Regime

To date, international research collaboration has worked with a minimum of rules and regulations. States apply phyto-sanitary controls over plant material entering their territories, and enforce quarantine regulations which have slowed or impeded the flow of research material.

Nonetheless, little control has been exercised over genetic information embedded in such material, and over its exchange. In the past, collection of genetic material by individual scientists, botanical gardens and gene banks from foreign countries attracted little attention from government agencies. Where it did, and where approval and/or special entry visas were required, host countries generally permitted collection missions on condition that their own scientists participated and samples of collected material were provided for storage at a local facility.

Intellectual property protection was of little concern. In the genetics field, it still is a comparatively recent phenomenon, which started with the introduction of plant variety protection in some industrial countries in the 1930s, and was more recently extended by allowing regular (utility) patent protection to life forms.

To date, international research collaboration has worked well under the free-exchange regime. An example of successful international research cooperation has been the development of high-yielding wheat and rice varieties by CIMMYT and IRRI[3] in the 1960s. The Green Revolution brought together the best brains in plant breeding technology and knowledge to accomplish, without a single patent or plant breeders' right, one of the great technological breakthroughs of modern days. No less importantly, its breeders were able to draw on germplasm resources from all parts of the world, and the systematic utilization of global genetic resources led to the success of the Green Revolution.[4]

Because the use of plant genetic resources has been reviewed as central to their crop improvement programs, the international research centers grouped in the Consultative Group on International Agricultural Research (CGIAR)[5] have subscribed to a policy of free exchange of germplasm.[6] Germplasm from their global collections of genetic material of the major food crops is freely available to scientists anywhere in the world, public or private, who can use it without restrictions for research or commercially. This has been the rule in their relations with partner institutions in both developing countries where the Centers collaborate with the national research systems, chiefly in the public sector, through which they disseminate their research products, and in industrial countries where they draw on advanced laboratories in industry and universities for inputs into their research programs.

The rule of free exchange is still observed by major genebanks in the industrial world. As recently as 1990, the U.S. law established that genetic material assembled by its National Genetic Resources Program must be freely available upon request.[7] From its 400,000 accessions at the U.S. National Seed Storage Laboratory, some 200,000 samples were sent out in 1991 of which about 20 percent had been requested by researchers resident in developing countries.[8]

If It Ain't Broke, Why Fix It?

Although it has served the interests of the international research community well, the free exchange system is under siege. Current pressure comes from the developing countries. There are several reasons for this. First, the advent of biotechnology has led to broad acceptance of intellectual property protection of genetically manipulated plant material. Developing countries are concerned that genetic material originating from their territories, once improved through breeding and research in industrial countries, will be protected as intellectual property, and no longer be freely available to them. Because they provided the original input on which improvements have been built, they argue that they be given a share in the commercial profits. Second, developing countries are realizing the importance of unexplored germplasm as an input into modern breeding programs, particularly since biotechnology has made possible the transfer of genetic traits across species. At the same time, these countries continue to depend on technologies generated by developed countries. In this sense, developing countries have discovered that they are holding within their territories a bargaining chip in the form of raw germplasm which represents the majority of the worlds's untapped resources. Third, the free exchange system gave developing countries the finished varieties to grow the food to feed their increasing populations, but did not succeed in giving them the technology to enhance their own seed material.

These concerns have validity. Under intellectual property protection, research collaborations will be different. Research information will flow less freely. In the laboratory and at the greenhouse bench, record keeping requirements will be more exacting. Publication may be delayed until the potential for commercial benefits has been assessed which would justify intellectual property protection. Yet, considering the cost of intellectual property protection, only a fraction of innovations will in the end be so protected.

Price effects of intellectual property protection on the cost of developing country seed requirements are difficult to assess. While improved seed will cost more than traditional seed, farmers in developing countries will be ready to pay as long as the price differential is justified by higher yields or other traits for which a farmer may buy it. What some have interpreted as a disappointing rate of technology transfer to developing countries cannot be ascribed to deficiencies in the free exchange system, but rather to the absence of an adequate research infrastructure in many developing countries. For both transfer and dissemination of technology in agriculture, developing countries rely on their national public research systems such as government research services and local universities. Because there is little or no private agricultural R&D in developing countries, the public systems are the only conduit available to adapt research products available internationally and to make them available to local farmers. Yet, because developing countries generally have given insufficient priority to public agricultural research and have provided inadequate funding and staffing for it, their existing research infrastructure has been less than effective in discharging its role in the transfer of technology.

Some argue that if developing countries provided effective protection of intellectual property, more effective dissemination of research products in the developing world would be encouraged. However, a number of other forces must be recognized. First, even when knowledge advances are broadly disseminated (including improved germplasm), their utilization may be disappointing. This may be the case where investment in new products and processes the utilization is discouraged when those innovations can be freely produced or copied, in the absence of an effective legal framework for protection of intellectual property embodied in the innovations. While it is difficult to judge whether intellectual property protection would encourage germplasm dissemination, one should expect that it would kindle private sector breeding activity in developing countries. Thus, except for the more advanced developing countries, one should not expect immediate positive effects for research dissemination and technology transfer from stronger protection of intellectual property in international research collaboration. To conclude, the need for renouncing the proven rule of free exchange of germplasm in international research cannot be demonstrated. Yet, political pressures may lead to its abandonment.

Towards A New Set of Rules in International Research Collaboration

Efforts have been underway for some time to internationally codify the access to, and use of, plant genetic resources. In 1983, the Food and Agriculture Organization of the United Nations (FAO) established a set of rules known as the "International Undertaking on Plant Genetic Resources".[9] Essentially, it is an attempt to stop or slow the rapid and uncontrolled disappearance of crop plant species from genetic erosion through international conservation efforts: developing countries would take charge of conservation of genetic material in their territories; and industrial countries would meet the cost of conservation.

The International Undertaking originally subscribed to the rule of free exchange. To quote from its Article 1: "This Undertaking is based on the universally accepted principle that plant genetic resources are a heritage of mankind and consequently should be available without restriction".[10] Nonetheless, disagreement over the scope of intellectual property protection, and specifically over whether breeder's lines and material protected by plant breeders' rights should be available without restriction, led to the subsequent narrowing of the free exchange principle. In 1989, the FAO adopted two resolutions[11] providing an "agreed interpretation" according to which plant breeders' rights are not incompatible with the Undertaking, meaning that such materials remained outside the Undertaking.

In exchange for this concession to industrial countries, developing countries won endorsement of the concept of farmers' rights.[12] Farmers' rights are not rights in a legal sense, but a moral commitment given by the industrial countries to recognize and reward "the enormous contribution that farmers of all regions have made to the conservation and development of plant genetic resources." The FAO has not yet agreed how the farmers' rights commitment will be translated into action except that payments should be directed into an FAO Fund for Plant Genetic Resources. The concept of farmers' rights and its acceptance in the FAO, though not yet consummated, marks the first time that a *quid pro quo* rationale has been introduced into the discussion on access to germplasm. Accordingly, developed countries will owe more for its use than payment of the cost of its conservation.[13] The free-flow principle was further restricted though not yet formally abandoned when in 1991 FAO adopted another resolution[14] which still recognized the common heritage principle, but subordinates it "to the sovereignty of the states over their plant genetic resources." It effectively abrogates the free availability rule by declaring that "conditions of access to plant genetic resources need further clarification." Thus, the credo of the common heritage of mankind, though deprived of much of its practical content, formally survived in the FAO. Despite this tradition, it found no place in the Biodiversity Convention.

International Research Collaboration under the Aegis of the Biodiversity Convention: An Assessment

The Convention on Biological Diversity is one of five documents adopted by the Earth Summit in Rio de Janeiro in June 1992. The Rio Conference represented the first major international attempt to establish a blueprint of actions to combat a broad range of environmental problems, including global warming, the destruction of tropical forests, and the need to preserve the world's biological resources on which the survival of mankind depends. More than 150 nations attending the Summit signed the conference documents. The

U.S., while signing the other four documents, did not sign the Biodiversity Convention.[15]

The key features of the Convention on Biological Diversity can be summarized as:

- The Convention recognizes the intrinsic value of biological diversity of all life forms, and the critical importance of conservation and sustainable use of plant genetic resources "for meeting the food, health and other needs of the growing world population" (preamble).

- The Convention requires its contracting parties, developed and developing, to conserve and manage their biological resources. It establishes specific obligations to identify these resources, monitor their status and conserve them, both *in situ*, in their natural habitat, and *ex situ*, in genebanks and other storage facilities (articles 7 and 8). Members will report periodically on their actions and the effectiveness of their actions to a conference of contracting parties to be held at regular intervals (article 26).

- The Convention formally recognizes the sovereign control by individual nations over biological resources on their territories.

- The Convention requires industrial countries to allow and facilitate access to their technologies on mutually agreed terms, but recognizes the primacy of intellectual property protection as the limiting factor in any such release of technology: "In the case of technology subject to patents and other intellectual property rights, such access and transfer shall be provided on terms which recognize and are consistent with the adequate and effective protection of intellectual property rights" (Article 16.2). This principle is reinforced in article 19.2 which stipulates with "priority access" for developing countries to the results of biotechnology innovations (but not to the innovations themselves) developed from their genetic resources; it again requires that this be done on mutually agreed terms. There are no provisions for compulsory licensing.

- A key element in the Convention is a financial mechanism to be subscribed to primarily by the developed countries. Its objective is to fund developing country expenses on conservation as well as their access to technology. It could be drawn upon to pay for royalties.

• Article 15.3 excludes from the Convention genetic material collected prior to coming into force of the convention, i.e. all genetic material collected and currently stored outside its country of origin.

The Biodiversity Convention is less than a shining example of international rule making. It is fuzzy in many places and inconsistent or contradictory in others. Yet, few would disagree with the basic philosophy of the Convention. If conservation of biodiversity is for the common good, all should share in the burden according to their capacity. The Convention attempts to balance the interests of the South and the North: essentially, developing countries agree to conserve their genetic material and make it available in exchange for developed country financing of the cost of conservation, and for allowing developing countries access to the relevant technology.

The conservation commitments in the Convention are comprehensive, requiring countries to develop conservation strategies to identify, monitor and manage biological resources. The Convention contains detailed provisions on the establishment and management of protected areas (*in situ conservation*) and genebank (*ex situ*) storage, on training and public awareness measures, and requires environmental impact assessments on projects that adversely affect biodiversity. Because each commitment is prefaced by the words "as far as possible and as appropriate", they are no more than best-efforts commitments, and thus weak.[16] But so are all other commitments under the Convention, whether they pertain to developing or developed countries.

The Convention establishes its own financial mechanism which is to receive regular contributions. Nonetheless, no agreement has been reached on its governance: developing countries argued for a new fund under majority control, while industrial countries insisted that funding should be provided from the Global Environment Facility (GEF) jointly administered by the United Nations Development Programme, the United Nations Environment Programme and the World Bank. The latter proposal was adopted as an interim solution until the members of the Convention can decide on another arrangement (Article 39). Some countries have pledged contributions to the GEF in support of the Rio agreements, and since industrial countries are unlikely to contribute to a funding arrangement that allocates fund on a one-country-one-vote basis as developing countries had advocated, the interim arrangement may turn out to be fairly durable.

New and additional funding is to be provided by developed countries to enable developing countries to meet the full incremental cost of their obligations under the Convention. And, as Article 21.4 makes clear, only to the extent that developed countries meet this obligation will developing countries live up to their commitments under the Convention. Developing countries will not take a financial stake in the implementation of the Convention. Article 21.1 can be interpreted as allowing mandatory assessments of contributions from industrial countries "as decided periodically by the Conference of the Parties." These countries vehemently objected to efforts to introduce mandatory contributions, and accordingly, provided their own interpretations at the time of signing.[17]

The Convention specifies that technology related to conservation and use of genetic resources should be made available to developing countries; in regard to biotechnology it limits such access in two ways: to the country which has provided the genetic material; and to the results and benefits arising from biotechnologies, not proprietary technologies themselves. The terms have to be mutually agreed. This broadly protects the interests of a patent holder. The concerns of some quarters of the U.S. biotechnology industry which seems to have driven the U.S. refusal to sign the Convention in Rio, were essentially twofold: an implicit risk of compulsory licensing; and loss of progress made in protecting intellectual property in other international negotiations.[18]

As the name indicates, compulsory licensing allows a country to browbeat a foreign patent owner into licensing his/her invention to a local producer on grounds of national economic or security interests. Compulsory licensing is accepted under the Paris Convention on Patents and Trademarks (the century-old treaty which regulates the mutual protection of these two types of intellectual property rights). It is rarely practiced, not even by developing countries who, however, are its strongest proponents.

In the current GATT negotiations (the Uruguay Round) which if completed successfully may set international minimum standards for intellectual property protection, the U.S. fought hard to ban the concept of compulsory licensing (stealing by another name as one U.S. Trade Representative liked to call it); until it was pointed out that the concept is known and applied in the U.S. defense complex; the right of eminent domain does allow the use of protected technology for national security reasons. Consequently, in the latest GATT draft, compulsory licensing and eminent domain are treated identically under a heading "Other Use without Authorization of the Rights Holder" and both permitted under certain conditions. One should expect the U.S. to accept these provisions as part of an overall "package deal" in the Uruguay Round.

As pointed out earlier, the Biodiversity Convention makes no explicit provision for compulsory licensing. Concerns that it does so implicitly, center on articles 16 and 19 on developing access to technology. However, as earlier noted, such access is predicated on mutual agreement, thus precluding a license which would force a patent owner to cede the protected innovation against her will. Once an owner declines a request for transfer this will end negotiations. Even the owner's government cannot force the patent owner to make the innovation available.

After the U.S. quietly abandoned its campaign against compulsory licensing in the GATT negotiations, apprehensions that the provisions of the Biodiversity Convention could weaken, or that erosion could occur in progress made in those negotiations with regard to the protection of intellectual property seem misplaced.

A very serious concern is that the Biodiversity Convention could slow or impede the flow of genetic material, and with it, international research collaboration. Article 3 grants States "the sovereign right to exploit their own resources according to their own environmental policies," while Article 15 states: "Recognizing the sovereign rights of States over their national resources, the authority to determine access to genetic resources rests with the national governments and is subject to national legislation." This terminology goes well beyond the letter of the FAO Undertaking, even in the interpretation of the 1991 FAO Resolution, in that it denies a commitment to release and exchange germplasm. Article 15.2 merely requires a state to "endeavor to create conditions to facilitate access to genetic resources for environmentally sound uses ... and not to impose restrictions that run counter to the objectives of [the] Convention".

Some users of germplasm may be less concerned because (a) genetic material they need is already stored in genebanks and thus removed from the scope of the Convention; and (b) what is not in genebanks can probably be obtained outside the convention; after all, the export of genetic material is difficult, if not impossible to police; a few seed grains are normally enough to export a plant variety. Yet, many users seek systematic access to resources not yet collected and stored, e.g. the agricultural biotechnology industry.

The user community, following their different interests, is likely to react differently to the new situation. Industry will swiftly adjust to the new situation. Merck & Co., the largest U.S. pharmaceutical manufacturer, recently (and well before the approval of the Biodiversity Convention) set an example: For payment of $1 million and provision of field equipment, Costa Rica's Instituto Nacional de Bioversidad will provide an unspecified number of probes (plants, insects, microbes) at Merck's choice over a period of two years. Merck has the right to patent inventions it makes from Costa Rican material, but has promised to pay INBio an undisclosed percentage of sales revenues. INBio intends to use the proceed in its on-going biodiversity conservation programs.

In their search for genes, agricultural biotechnology industries are likely to follow the lead of the pharmaceutical industry, holding out smaller monetary rewards coupled with access to improved seed material and collaboration agreements. The scope for joint research arrangements seems large in the agricultural area, and certainly greater than in the pharmaceutical field; while the odds of finding a blockbuster product are smaller.

Academic and other public research groups may not be able to compete on the terms that industry can offer, especially with up-front payments. They may have other options: they can form alliances with industry, as Cornell University did in the Merck-Costa Rica deal. Or they may offer research and training services in exchange for genetic material. The arrangement which the National Cancer Institute offers seems to point in this direction. The NCI annually procures some 6000 samples of plants, marine organisms and microbes from developing countries. In exchange it trains scientists from source countries and promises to negotiate a share of royalties on behalf of the source country with the eventual manufacturer of a cancer drug.

Public genebanks, the major players in the current campaign to save the world's genetic resources, may not be able to do this. If a genebank cannot pay, and has little else to offer in exchange for genetic material[19], developing countries may release genetic material only against a commitment from the genebank to restrict the use and distribution of the material. Supplying countries may, for instance, request a genebank to track every subsequent user who then would be required to share royalties with the source country. As genebanks have generally no control over distribution except for their first release, they may not be able to enter into such commitment; which, in turn, could seriously impede their collection efforts. The international agricultural research centers which hold one-third of the world's food crop germplasm in their genebanks, are in a similar situation, except that because of their location in, and close relations with, developing countries they have a larger scope for collaborating with these countries, and in exchange maintain open, or at least easier, access to germplasm resources.

It would be ironic if the mission of public genebanks in industrial countries, who have demonstrably worked for the benefit of all germplasm users including the developing countries, is hampered by the Biodiversity Convention, while industry, especially the pharmaceutical and biotechnology industries, which in North-South polemics frequently appear as the archenemy of developing country aspirations, would retain access.

This may well be the most difficult problem to solve in the implementation of the Convention. An arrangement will have to be found which recognizes the interests of developing countries in securing some reward from the genetic material they provide, while accepting the essential role public genebanks play for the common good which should not be burdened with comprehensive tracking responsibilities for the germplasm they release. The answer may lie in an honors system under which users of genebank material agree to voluntarily pay a reward to the source country in case of successful commercialization of a product derived from that country's genetic material.

Public genebanks in industrial countries should also form their own alliances. The prospect of sharing technical know-how such as storage techniques and database management should make it attractive for developing country genebanks to join. Such alliance should establish their own free-exchange and release rules. The international centers which to date have sent out probes without formality to any *bona fide* researcher worldwide, in future may have to do so under material transfer agreements which then allow a center to claim a share of the profit from commercial use. This could take the form of a royalty share or a license granted to the center for use by a center in research work for developing countries. Alternatively, royalties could be passed to source countries if important enough and unless they would have to be divided among too may providers of parental material.

Conclusions

For the first time, nations have undertaken broad commitments to preserve biological diversity. While specific commitments are predicated on what is "possible and appropriate", establishment and implementation of conservation programs in individual countries can in future be effectively monitored. Like the conservation commitments, financial commitments under the Convention are moral in nature, and not legally enforceable. Like in all other international funding arrangements, peer pressure will be required to ensure that total funding needs are met.

Commitments to transfer environmental and conservation technologies to developing countries are unlikely to present a burden to industries from the developed world. The patent rights of the biotechnology industry are not threatened by the Biodiversity Convention. For the U.S. not to sign the Convention over such concerns was almost certainly a mistake which is likely to be corrected by the new Administration.

The Convention could impede access to germplasm from the developing world. Industry is likely to adjust to the rules of the Convention by paying, or paying more, for access rights. Public research institutes will come under similar pressure, but should be able to offer compensation in form of services such as exchange of scientists and training commitments. This may foster collaboration between developing and developed research establishments. Public genebanks in industrial countries, however, could be seriously hurt in their efforts to salvage germplasm resources through *ex situ* storage. Twinning and network arrangements may offer possibilities to help them continue their missions. Similarly, for the international research centers in the CGIAR, it will be essential to involve developing country research agencies more effectively in their efforts. If developing countries can see themselves as full stakeholders in the CGIAR enterprise, they are more likely to continue the free supply of germplasm to the centers.

The Convention provides a rough framework for international collaboration in the conservation of biological diversity. Protocols remain to be negotiated between countries on access to genetic resources and technology transfer. Its success will depend on the spirit and attitude with which countries pursue its implementation. If this takes place in an adversarial spirit, as negotiations of the Convention in Nairobi and Rio de Janeiro were unfortunately concluded, the consequences would hurt developing and developed countries alike. If, however, all parties convince themselves that they are partners in this venture, and that their mutual interests are best served by collaborating in implementing the Convention, concerns about restrictions on the flow of genetic material should prove invalid. The challenge to feed the world's growing population within a sustainable and liveable environment hardly leaves a choice.

Endnotes

1. Resolution 8/83 of the Twenty-second Sessions of the FAO Conference, Rome, 5-23 November 1983.

2. UNEP "Convention on Biological Diversity".

3. Centro Internacional de Mejoramiento de Maiz y Trigo (Mexico), and International Rice Research Institute (Philippines).

4. For a review of the results of the Green Revolution see Anderson et al.

5. Established in 1971, the CGIAR is an association of countries, international and regional organizations, and private foundations dedicated to supporting a system of agricultural research centers and programs around the world. Eighteen centers currently belong to the Group.

6. TAC Document AGR/TAC:IAR/88/4 "CGIAR Policy on Plant Genetic Resources", Rome, February 1988

7. Sec. 1632(a)(4) of Public Law 101-624 of November 28, 1990 reads: "The Secretary [of Agriculture], ...shall (4) make available upon request, without charge and without regard to the country from which such request originates, the genetic material which the program assembles."

8. Communication from Dr. Steve Eberhart, U.S. National Plant Germplasm System.

9. Resolution 8/83 of the Twenty-second Session of the FAO Conference, Rome, 5-23 November 1983. In 1990, the U.S. joined the FAO Commission on Plant Genetic Resources which administers the Undertaking, but has not signed the Undertaking.

10. Article 5 of the Undertaking specifies that "[i]t will be the policy of adhering Governments and institutions having plant genetic resources under their control to allow access to samples of such resources, and to permit their export, where the resources have been requested for the purposes of scientific research, plant breeding or genetic resource conservation."

11. Resolutions 4/89 and 5/89 adopted by the Twenty-fifth Session of the FAO Conference, Rome, 11-29 November, 1989 and incorporated into the International Undertaking as Annexes I and II, respectively.

12. "Farmers' rights mean rights arising from the past, present and future contributions of farmers in conserving, improving, and making available plant genetic resources, particularly those in the centers of origin/diversity. These rights are vested in the International Community, as trustee for present and future generations of farmers ..." (Resolutions 5/89 ibid).

13. It should be noted that the farmers' rights concept, by definition, only covers genetic material in the hands and on the land of farmers, and not its weedy relatives which are of increasing importance in biotechnology research.

14. Resolution 3/91 adopted by the Twenty-sixth Session of the FAO Conference, Rome, 9-28 November 1991 (now Annex 3 to the International Undertaking).

15. For an account of the circumstances that led the U.S. president to decline signature of the Biodiversity Convention, see Gore (Foreword to the Plume Edition); hearings of the U.S. Senate Foreign Relations Committee on September 18, 1992; and Porter.

16. Barton points out that a number of more specific commitments were removed during the final round of negotiations.

17. The United Kingdom declared that "nothing in Article 20 or Article 21 authorizes the Conference of the Parties to take decisions concerning the amount, nature, frequency or size of the contributions of the Parties under the Convention." France, Italy and Switzerland issued similar declarations.

18. Porter 10-12.

19. They could offer their better equipped storage facilities to genebanks from developing countries. However, distrust in North-South relations has sofar not permitted such arrangements on a larger scale.

References

Anderson, Jock R.; R. W. Herdt; and G. M. Scobie. 1988. *Science and Food: The CGIAR and its Partners.* World Bank, Washington, D.C. 1988.

Barton, John H. 1992. *Biodiversity at Rio.* BioScience November 1992

Technical Advisory Committee of the CGIAR (TAC). (February) 1988. *"CGIAR Policy on Plant Genetic Resources."* Document AGR/TAC:IAR/88/4. Rome, February 1988

Gore, Albert. 1992. *Earth in the Balance: Ecology and the Human Spirit.* Plume, New York, N.Y.

Porter, Gareth. 1992. *The United States and the Biodiversity Convention: The Case for Participation.* ESSI Papers on Environment and Development Number 1. Environmental and Energy Study Institute, Washington, D.C..

United Nations Environment Programme (UNEP). (June) 1992. *Convention on Biological Diversity.* Doc. Na.92-7807.

United Nations Environment Programme (UNEP). *Conference for the Adoption of the Agreed Text of the Convention on Biological Diversity.* Nairobi Final Act, Doc. Na.92-8138.

United Nations Food and Agriculture Organization (FAO). 1983. Report of the Conference of FAO, Twenty-second Session. Rome.

United Nations Food and Agriculture Organization (FAO). 1989. Report of the Conference of FAO, Twenty-fifth Session. Rome.

United Nations Food and Agriculture Organization (FAO). 1991. Report of the Conference of FAO, Twenty-sixth Session. Rome.

United States Senate Foreign Relations Committee. (September) 1992. *UN Framework Convention on Climate Change.* Federal News Service.

Strategies for Public Sector Agricultural Research

Chapter 16

Changes in Agriculture and Agricultural Institutions

Donald Holt, Director
Illinois Agricultural Experiment Station

This paper is about land-grant colleges of agriculture within land-grant institutions and their relationship to the industry of agriculture. The general ideas apply to other agricultural institutions and agencies as well. The research arms of land-grant colleges of agriculture are collectively designated State Agricultural Experiment Stations (SAES). The Cooperative Extension Service (CES) provides outreach for these institutions. The United States Department of Agriculture-Agricultural Research Service (USDA-ARS), part of which is located at land-grant institutions, performs a closely related research role. Other USDA sub-agencies likewise perform important research and/or educational roles. These are all important components of the institutional structure of U. S. agriculture.

Historic legislation, including the Morrill (1862), Hatch (1887), and Smith-Lever (1914) Acts, created land-grant institutions. That legislation provides a clear mandate for those institutions to serve: the industry of agriculture, broadly defined; the people who work in the industry; and the consumers of agricultural products and services. That service is to include relevant instruction, research, and extension.

In accord with the design principle that "form follows function," land-grant colleges of agriculture should be structured and managed so that a close relationship exists between the institutions and the agriculture they serve. In fact, they should be an important part of that industry. The institutions were and still are architects of change in agriculture. Though other economic, social, and political forces change agriculture. Agricultural institutions should be flexible enough to adjust accordingly. Of course, if certain changes in agriculture turn out to be undesirable, the institutions should help the industry make appropriate adjustments.

This is a time of uncertainty and reflection in agriculture. People in all parts of agriculture are seeking a shared vision of mission and objectives (Holt, 1992a). Some suggest that the institutions are "adrift" (Wheeler, 1992). There is concern about changes that have occurred and whether or not they are good. Questions are raised as to whether agricultural institutions have adjusted to changes in the industry and its technological, economic, and social environment. High-level decision-makers are asking how agriculture and its important institutions should be configured to cope with the agricultural situation of the future.

Alternative Perspectives on the Land-Grant College of Agriculture

Land-grant colleges of agriculture have come under considerable scrutiny over the last two decades. Reports of the 1970's and 1980's (cited by Fox et al., 1987) criticized agricultural research institutions and agencies for failing to employ basic biological science to address agricultural problems and opportunities. Such groups argue that building the basic agricultural knowledge base is the best approach (Board on Agriculture, 1989). This is to be accomplished by providing more support to narrowly focused, disciplinary specialists in the sciences basic to agriculture. They propose involving more basic scientists, including those in non-agricultural institutions, in agricultural research.

In a sense, these people are arguing for a more reductionist approach to solving agricultural problems. Members of the basic science community focus on the specific components of agricultural systems, seeking to understand and improve those components. While they foster interdisciplinary basic research, they often rely on others to assemble and integrate specific improved components into effective, practical agricultural systems.

To illustrate, basic scientists may learn how a specific genetic control mechanism works. They may even incorporate an improved version of that mechanism into a plant. They probably will not be directly involved, however, in creating a new commercial crop variety with that mechanism or testing it in the many soil, climatic, and socioeconomic environments in which it will have to perform. They probably will not work on the new crop management practices or other new technologies by which the full potential of genetic change can be realized.

The National Initiative for Research on Agriculture, Food, and the Environment (NRI), a major competitive grants program administered by the USDA-Cooperative States Research Service (USDA-CSRS), grew out of the perceived need for more basic research. It was proposed and shepherded through the political process by the Board on Agriculture of the National Research Council of the National Academy of Sciences. Now the Board on Agriculture is undertaking a broad study of public agricultural institutions to determine how well they are adapting to changing agricultural conditions. It would not be surprising if those who conduct that study approach it from the viewpoint of basic science and reductionism.

Various groups associated with the sustainable agriculture movement criticize agricultural institutions for contributing to the development of capital-intensive agriculture. They see this as causing various environmental and natural resource problems; reducing the number of small, family farms; and causing the decline of rural communities. Such groups attribute some agricultural problems to a reductionist approach to agricultural science and technology. They argue for more holistic, systems approaches to agricultural problems and opportunities. They would like to see stronger connections between publicly supported agricultural research programs and the practical outcomes they desire.

Sustainable agriculture groups suggest that farms should be more diversified and that there should be less division of management and labor in agriculture. They think farmers should rely less on purchased inputs and more on inputs that can be generated on the farm, e.g., manure and legumes in place of commercial fertilizer. The history and philosophy of the sustainable agriculture movement is discussed by Harwood (1990).

The Sustainable Agriculture Research and Education Program (SARE, formerly known as LISA, i.e., Low-input Sustainable Agriculture) was designed to address sustainability concerns. This regionally administered program provides grants for farm-level, systems research involving coalitions of farmers, agricultural scientists and educators, and various non-profit organizations associated with sustainability issues.

Another broad study of agricultural institutions is being sponsored by the Kellogg Foundation. Judging by the composition of the steering committee of that study, the Kellogg study will likely approach the evaluation of agricultural institutions from a sustainable agriculture perspective.

Influential members of Congress have called for a major reorganization of USDA. Some proposals for reorganization would affect science and education programs. This is further evidence of the scrutiny under which agricultural institutions now find themselves.

Behind each of these proposed studies and reorganization plans is at least an implication that public agricultural institutions have not responded to dramatic changes in agriculture and that they need to be rather extensively reconfigured. There seems to be considerable difference of opinion, however, on how agriculture has changed and how agricultural institutions should change to play their role most effectively.

Agriculture Then and Now

Until the last half of the nineteenth century, U. S. farms were essentially self-sufficient. Crops, including grain, vegetable, and fruit crops, and livestock were produced, processed into food, fuel, clothing, and shelter, and consumed on the farm. Farmers hauled some produce to town and marketed it directly to their city neighbors, who comprised a relatively small proportion of the population. Most inputs, especially labor, were generated on the farm. Farmers made some of their own equipment. Some made and repaired equipment for their neighbors. Farmers used manure produced on the farm for fertilizer. They used hoes to control weeds. They were often the source of capital for their neighbors. They were also a source of agricultural information.

Meetings and publications of such organizations as Farmers' Institutes covered a broad range of topics related to agriculture. They provided information on how to produce crops and livestock, how to process agricultural produce, and how to prepare it for meals. Farmers and rural homemakers were the source of much of this information.

In short, in earlier times farmers were the producers, processors, distributors, retailers, and consumers of agricultural products and performed almost all of the support functions.

U. S. agriculture still accomplishes the same things it did 100 years ago. Agricultural products are produced, processed, distributed, retailed, and utilized. The major difference is that much of that activity no longer occurs on farms. There has been a tremendous division of labor and a great degree of specialization within agriculture. A huge infrastructure of financial institutions, suppliers, equippers, builders, transporters, farm and trade organizations, consultants, government agencies, colleges of agriculture, other institutions, and media groups accomplishes what was once accomplished almost entirely by farmers.

Many farmers rent land for farming, thus separating ownership and operation. Farmers usually purchase seed rather than producing it on the farm. Farmers plant crops but often hire private firms to apply fertilizer and pesticide. Equipment may be leased and professional mechanics employed to repair and maintain it. Consultants may be hired to help make some or all management decisions. Currently, the typical American farmer is a producer of raw materials. Others use those raw materials to produce and market finished agricultural products and services. Similar changes have occurred with regard to fisheries and forests, which are important components of modern agriculture.

The major external force driving this change is the logistical challenge of feeding increasingly large urban populations. As these urban populations become more affluent, they demand better quality, safer, more affordable, and more convenient agricultural products and services. Agriculture must serve the basic needs of a largely urban population using a relatively small workforce. That workforce is rapidly becoming less inclined to the labor-intensive agricultural methods of the past.

Agricultural activities continue to become more highly specialized, because specialization provides many advantages. A generalist may be able to do a specific task, but one who specializes in that task probably can do it better. A division of labor permits specialization, which improves the way specific tasks, including management, are accomplished. Specialists and specialized firms provide their special services and/or produce their special products for many clients and customers. In this way they achieve great economies of scale. They pass some of these economies on to their customers in the form of lower prices. High-volume specialists also have considerable influence over their suppliers and are able to negotiate lower prices on the things they need to purchase. Part of these savings also are passed on. High volume specialists also are able to serve more than one client group. This helps to spread their business risk.

The division of labor within U. S. agriculture tremendously increased its efficiency. Most of the benefits accrued to consumers of agricultural products. The agricultural workforce decreased in size to a little over 20 percent of the population. This freed almost 80 percent of the nation's human and other resources to provide and accomplish all the other things that contribute to a high standard of living and a great civilization.

This stands in stark contrast to, say, the People's Republic of China. The unspecialized agriculture of China, in which farmers do everything, requires 60 to 80 percent of the nation's workforce. Most of the developing nations have a similar problem.

Land-grant institutions responded to changes in agriculture by developing more specialties. Agriculture departments became colleges of agriculture made up of departments of agronomy, forestry, animal sciences, agricultural engineering, and agricultural economics. Subsequently, they further subdivided into a number of other agricultural disciplines and subdisciplines, e.g., food science, foods and nutrition, horticulture, plant pathology, entomology, plant breeding and genetics, textiles and apparel, consumer and family economics, and others.

These more specialized departments, disciplines, and subdisciplines were organized around specialized client groups within commercial agriculture. For example, at the University of Illinois, we have a strong agricultural finance research program that serves financial institutions and financial policy decision-makers. We conduct research and training in agricultural communications, thus serving media groups.

Land-grant colleges of agriculture mount research, teaching, and extension programs serving every component of the industry of agriculture including several different consumer-related groups. They also conduct agricultural research addressing the concerns and interests of environmental, conservation, and natural resource groups. In fact, the focus on environment and natural resources is engendering a new agricultural discipline.

Problems Caused by Specialization

Traditionally, agricultural institutions, especially land-grant colleges of agriculture, were identified with farmers and farms. On university campuses, reference to the "cow college" was not unusual. This was appropriate when almost all of agriculture was on farms and almost all clients and constituents of agricultural colleges were members of farm families. As the various components and activities of agriculture and agricultural colleges became differentiated and specialized, the public perception of agriculture and its institutions did not change. The word agriculture still connotes farming and farmers to many people. Ironically, the further people are removed from their agricultural heritage, the more they cling to this old-fashioned concept of agriculture.

There are some good aspects of the general public's warm feelings about the family farm and the farm family. Agricultural institutions should capitalize on that by acknowledging the rich cultural heritage of American agriculture and manifesting the best of its cultural attributes. On the other hand, many citizens, including some influential decision-makers, note that farmers now make up a small proportion of the population. They conclude, erroneously, that agriculture and agricultural institutions are no longer as important as they were in the past.

People employed in agricultural institutions are themselves specialists who are increasingly removed from the farm. One of the biggest challenges facing agricultural institutions is educating their own people, the general public, and key decision-makers on the enormous scope and complexity of agriculture.

People need to understand the fundamental role of agriculture in meeting basic human needs and as the foundation of the nation's economy. They also need to be aware of the global interdependence of nations with regard to agriculture. They need to appreciate the importance of strong, well-supported, and well-managed agricultural research and education institutions.

Growing even more rapidly than the world population are the expectations of the world's people for quality of life. The powerful forces of population and economic growth will impose an enormous burden on the world's agricultural systems and the natural resources used by those systems. Public agricultural institutions will have to play key roles in meeting these challenges. The situation provides unprecedented opportunity for valuable public service.

As the client groups of agricultural institutions become more differentiated and specialized, they require more sophisticated and specialized human resources, technology, and information. They also need more rapid response to problems and opportunities. To serve these groups, agricultural institutions have to spread their resources over programs of ever-increasing scope.

In such situations, it becomes extremely important for the institutions to remain focused on key agricultural issues and subject matter. Otherwise, resources are spread too thinly and programs lose quality. If institutions expand their programs into subject matter covered by other public institutions and agencies, they lose their uniqueness and encounter competition from other specialists, some of whom may be better qualified to deal with some subject matter.

The end result of this problem may be that agricultural needs are not met and institutions are perceived as less effective. Currently, there is confusion and disagreement as to the appropriate scope for the research, teaching, and extension programs of land-grant colleges of agriculture (Holt, 1992a). The industry of agriculture still needs its agricultural institutions. They are the major information component of the industry, which is growing in complexity. In modern times, information is the principle competitive resource of any industry. The information component of an industry must grow in size and quality if the industry is to survive rigorous competition in the global economy.

Agricultural institutions need strong state, federal, and private support to adjust to the expanding scope of agricultural problems and opportunities. At the same time, they must maintain and cultivate a shared vision of mission and objectives and maintain their focus on agriculture in order to use public resources most effectively. As a college of agriculture engages and begins to serve new clients and constituencies, some of its older client groups may perceive that they have been abandoned or, at least, are not receiving the attention and quality of service they did before. Traditional client groups provide much of the political support for colleges of agriculture and other agricultural institutions. Lack of political support from clients and constituents can threaten the core support of agricultural institutions, that is, the regularly appropriated state and federal formula funds. Realistically, some group or groups that are well-organized and politically influential need to have a sense of ownership of agricultural institutions. Influential groups will no doubt expect to benefit from providing political support.

At least one other problem arises from specialization within agriculture. It impinges on all of agriculture. As agriculture and agricultural science is de-integrated into more and more specialties, it becomes more and more difficult to maintain adequate communication and coordination among these specialties. When that communication, coordination, and mutual understanding is missing, one component of agriculture may adopt practices or products that cause unexpected effects or problems somewhere else in the agricultural system. For example, off-site impacts of agricultural practices may be harmful. Serious problems may result, such as silting of reservoirs, contamination of water supplies, or fears about food quality. The regulatory reactions to these situations may themselves have unpredictable, possibly harmful, effects.

While specialization fosters excellent performance, it definitely is not advantageous to do something extremely well that you should not be doing in the first place. It takes considerable institutional and individual effort to maintain the level of communication and coordination needed to anticipate and prevent unpredictable results of specialization.

A primary concern of sustainable agriculture groups is that reductionist science and specialization cause people to lose sight of broader, system-related concerns. They see the agricultural system being reduced to pieces and the pieces being managed and modified individually. They fear that when the pieces are reassembled into the overall system, they may no longer fit.

Some individuals and groups feel there are too many conflicts of interest, particularly economic interest, for the individual, specialized components of agriculture to function for the common good. Others believe that problems caused by specialization are minimized as long as the focus is on serving the consumers of agricultural products and services. They believe also that if system-wide communication, coordination, and cooperation are fostered in the context of free markets, the system can and will function for the common good.

There is little doubt that the high degree of specialization within agriculture makes it harder to implement system-wide changes. Inability to adapt to change in the face of new situations is a symptom of overspecialization. Based on the experience of the dinosaurs, overspecialization is something to be avoided. Better communication and coordination is required to avoid overspecialization in agriculture.

Achieving adequate communication and coordination within agriculture will become an even greater challenge in the future as agriculture becomes even more complex and specialized. Manufacturing firms are already facing this problem and have devised various approaches to coping with it.

Manufacturers recognize that if they are to compete successfully in highly competitive global markets they must respond rapidly to changes in the needs and desires of their customers. There must be close communication and coordination among the business administration, product design and engineering, manufacturing design and engineering, production, marketing, accounting, shipping and receiving, and other divisions of each firm. In a very large firm, there may be hundreds of thousands of employees, representing many different skills and disciplines, scattered throughout the world in many locations. Maintaining the necessary level of communication and coordination is a daunting challenge. A modern approach to this problem is known as "computer-integrated manufacturing" (CIM).

CIM involves connecting all parts of the firm and almost all of its people in a computerized communications network. To illustrate how this works, all of the following can be accomplished, almost simultaneously, through the computerized communication system. A need for a design change is: identified by marketing/customer relations people; communicated to business administrators for approval; communicated to product design and engineering people, who incorporate the change into blueprints using computer-assisted design (CAD) software; automatically implemented on the production line via computer-controlled machines; and simultaneously communicated to marketing, accounting, and shipping people so they can change promotion and advertising materials, open and close accounts, develop shipping schedules, etc., as needed.

Achieving this level of communication and coordination in a single manufacturing firm is a great challenge even if the firm is under one roof, vertically integrated, and more or less self-sufficient. By considering cooperative efforts between two or more firms, this agriculture/manufacturing analogy can be taken to another level of complexity.

Just as each farm was once a vertically integrated agricultural operation, manufacturing firms were more self-sufficient in the past. Now, they achieve economies of scale and other advantages of specialization by out-sourcing many of the parts and components that they assemble into finished products.

When a manufacturing firm obtains specific parts or components from another firm, the other firm must deliver those parts and components in the quantities and on the schedule required for the manufacturing process. Also, the parts or components must fit in and be completely compatible with the final product.

To meet "time-based" competition and maintain the small inventories of "just-in-time" manufacturing, there must be extraordinary communication and coordination between manufacturing firm and supplier. In this situation, the challenge is to achieve excellent communication and coordination not only within a firm but also between firms.

As U.S. agriculture attempts to produce more differentiated products and export less produce in the form of raw agricultural materials, a similar coordination problem arises. For example, to produce one new processed meat product containing, say, lower levels of saturated fatty acids, may require that changes be implemented at several levels in a complex value-added process.

Research institutions, plant breeding firms, seed firms, grain producers, feed manufacturers, livestock breeders, livestock growers, livestock feeders, meat packers, meat processing firms, distributors and retailers of meat products, and consumers each may have to institute some change in their products and processes to produce, utilize, and capture the benefits of this single new development. Each such "value chain" has several converging and diverging value chains of inputs and co-products. If one of these "links in the value chain" leading to the new product is weak, that is, if farmers or firms accomplishing one of these "economic stages" do something wrong, the whole process may fail to meet the needs and desires of the consumer or the business goals of the participants.

Obviously, there must be excellent, continuing communication and coordination among the various people and firms involved in such complex, value-added activities. The rapid growth of contract production, processing, and marketing is a response to this need for greater communication and coordination within the industry of agriculture. Through contracts and other coordination mechanisms, "value-added partnerships" are formed and "vertical cooperation" is achieved. Increasingly, both intermediate and final agricultural products must be produced or manufactured to specifications.

Achieving adequate communication and coordination is a daunting challenge if only one manufacturing firm and its suppliers are involved. Think how much more challenging it will be to achieve it in the vast, enormously complex, geographically dispersed, and highly decentralized industry of agriculture.

Sonka (1991) discusses trends toward and mechanisms for better market coordination. Drawing on CIM as a model, (Holt, 1993) has described how modern information technology can be employed to create the "computer integrated agriculture" of the future.

"Time-Based" Agricultural Research and Development

If agricultural institutions are to play a useful role in the highly coordinated agriculture of the future, they must coordinate their research and education efforts with the efforts of the industry to serve the consumers of agricultural products and services and meet the needs of the agricultural workforce. This will require changes in the way these institutions organize their research and education efforts.

Agricultural R&D can be organized as a linear, stepwise process involving basic, developmental, and adaptive research, and technology transfer. Since basic research is organized in disciplines, disciplinary concerns and interests play a major role in the selection of subject matter for linear R&D.

In an alternative organizational scheme, the R&D functions are conducted simultaneously instead of sequentially. In this parallel approach, desired practical outcomes (goals) are identified and prioritized. These specific practical goals should be derived from and consistent with the mission and overall objectives of the sponsors of specific research programs.

In the parallel approach, the strategic question (What is the right thing to do?) is addressed before the tactical question (What is the right way to do it?). This is the logical planning sequence. Linear and parallel research efforts have also been described as "discipline-oriented" and "goal-oriented" R&D, respectively. The linear and parallel models of R&D organization are discussed in greater detail by Holt (1991).

In the parallel approach, once a specific practical goal has been specified, cross-functional and interdisciplinary teams can be organized to achieve the goals. The team effort also needs to be integrated over links in the associated value chains, including production, processing, distribution, and utilization of associated products. Those who must supply and implement the potential new information, technology, or regulations should be meaningfully involved in the planning and conduct of the R&D.

Business literature refers to the parallel model as "functionally integrated R&D" or "concurrent engineering." This approach is seen as superior to the linear approach as a means of achieving commercial (practical) goals. The time required to achieve specific goals is shortened with an accompanying decrease in the cost of the R&D program. Most important, firms that employ this organizational strategy beat their competition to the market with the best quality product or service. The parallel approach is particularly applicable in situations in which funds are limited and there is urgent need to achieve practical objectives quickly. Parallel R&D is absolutely essential in the age of time-based competition and just-in-time manufacturing.

The parallel model of agricultural R&D provides a conceptual framework for achieving the necessary degree of linkage between and integration of agricultural research and practice (Holt, 1992b). The desire of sustainable agriculture groups for more "mission-linked" research is addressed by the parallel model. Land-grant institutions have traditionally placed basic, developmental, and adaptive researchers and technology transfer people in the same administrative units. Usually individual scientists are responsible for more than one of these functions. This facilitates parallel R&D.

Modern agricultural institutions, especially land-grant colleges of agriculture, are in the best position to foster the required degree of communication and coordination within the industry of agriculture. Land-grant colleges of agriculture, perhaps more than any other component of agriculture, have relationships with all the other components.

These relationships have been established and cemented through the academic, research, and extension programs of the institutions. The institutions are the principle source of human capital, basic knowledge, and unbiased information and a major source of new technology for each of the many components of agriculture. They could also become an important mechanism of communication and coordination.

Greater coordination and communication within the industry will also resolve the problem of reductionist versus holistic approaches to agriculture. When there is easier communication and coordination among all the narrowly focused agricultural specialists, including agricultural scientists, each participant will better understand the problems and needs of the total system. They will be able to adjust their specialized activities accordingly. It will be in everyone's best interest to do so.

It will take a lot of institutional and individual energy to achieve and maintain the necessary level of communication and coordination within agriculture. Achieving it will be a great R&D challenge and an enormous technological achievement. If it can be achieved and maintained, however, agriculture and its customers and constituents can continue to reap the great benefits of agricultural specialization without its potentially damaging side-effects.

References

Board on Agriculture, NRC, NAS. 1989. "Investing in Research: A Proposal to Strengthen the Agricultural, Food and Environmental System." National Academy Press. Washington, DC.

Fox, G.; R. E. Evenson; and V. W. Ruttan. 1987. " Balancing Basic and Applied Science: The Case of Agricultural Research." Biosci. 37:507-509.

Harwood, Richard R. 1990. "A History of Sustainable Agriculture." In Sustainable Agricultural Systems. Clive A. Edwards, Rattan Lal, Patrick Madden, Robert H. Miller, and Gar House, eds. Ankeny, Iowa: Soil and Water Conservation Society.

Holt, D. A. 1991. " Organizational Paradigms of Agricultural Research and Development." Presented at the 1991 Annual Meeting of the Agricultural Research Institute, Bethesda, MD. (Copy available from Don Holt, Illinois Agricultural Experiment Station, 1301 W. Gregory Dr., Urbana, IL, 61801)

Holt, D. A. 1992a. "Achieving Focus in Agricultural Research and Extension." Proceedings of the Great Plains Agricultural Council. Lincoln, NE. June 9-11, 1992.

Holt, D. A. 1992b. Mission-linked research. Testimony submitted to U. S. Senate Subcommittee on Agricultural Research and General Legislation, The Honorable Thomas A. Daschle, Chair. September 17, 1992. (Copy available from Don Holt, Illinois Agricultural Experiment Station, 1301 W. Gregory Dr., Urbana, IL, 61801).

Holt, D. A. 1993. Designing and Marketing the Products of Computer-Integrated Agriculture. Proceedings of a Forum on Remote Sensing in Agricultural Decision-Making. GEOSAT, University of Oklahoma, Norman, Oklahoma, (in press).

Sonka, S. T. 1991. "New Issues in Agriculture: Concepts, Issues, and Opportunities." In: Paul A. Funk Recognition Program: Papers by Recipients of Awards in 1991. College of Agriculture, University of Illinois.

Wheeler, D. L. 1992. "Land-grant Universities Urged to Broaden Research Beyond Traditional Agricultural Mission." Chronicle of Higher Education, April 22, 1992, p. 48.

Chapter 17

Peer Review, Pork, and Priorities in Agricultural Research

Daryl E. Chubin
Office of Technology Assessment

Agricultural research is part of the *ad hoc* "federal research system" consisting of the R&D agencies, their constituent research communities and other interest groups, and the mass of executive and legislative branch structures that oversee the research fortunes of the nation. The system and its structural components are examined in the OTA report, *Federally Funded Research: Decisions for a Decade*. The clash of visions of agricultural research funding mechanisms and priorities, expressed both in policy and politics, plays out on this landscape. These visions can be seen, for example, in Fox et al. (1987), Hadwiger (1992), and Holt (1992).

As an analyst and congressional staffer, this author is an outsider to the agricultural research community. But I have long studied peer review, pork barreling (or the term I prefer, "earmarking"), and Federal science policy. In addition to OTA (1991) (as well as its predecessor, OTA (1990) these issues have been explored in Chubin and Hackett (1990). These studies and this paper are based on raw data including the numbers that the R&D agencies generate to characterize their sponsorship of research, as well as the words used by both research patrons and performers. It is important, then, that these data are viewed as claims to be evaluated with an open-minded, even-handed skepticism. Predictably, this sometimes disturbs both my congressional clients and their constituencies. This paper is offered in the same spirit.

The Issues

The chief issue confronting any R&D agency and its constituents today is not funding level, but the methods by which that funding is distributed: according to what criteria, with what expectations, in what time frame, etc.? Such decisions, of course, are not made in a vacuum. They are affected by national goals, a discretionary budget that reflects new initiatives, agency mission statements that stress ongoing programs, and the organizations that try to influence decisionmaking. The context for decisions, in short, is as important as the allocations that result from those decisions. One conclusion in this regard is clear, competing interests are rampant.

Anchored by the Department of Agriculture (USDA), the agricultural research system is a social invention with implications for U.S. industrial competitiveness. Some see USDA as a model of research funding:

It's time to move away from ideological posturing and think carefully about ways to organize tighter, more productive cooperation among government, the universities and industry. People who dismiss that as impossible might want to consider the example of agricultural research over the past century -- very possibly the most successful R&D program in the country's history -- and the relationship of the land-grant universities, the Agriculture Department and farmers working on their own land. The lessons of that immensely productive alliance seem increasingly applicable to American industry and its faltering progress in technology (Washington Post, 8/16/92).

But USDA is also cited, within the smaller world of federally funded R&D (in which this agency is the sixth largest source), as besieged by sometimes dated institutional and political considerations. For example, USDA-sponsored plant research has been charged by a National Research Council (NRC) committee as failing "to keep pace with major advances in molecular and cell biology . . . among others." The purported villain is the reliance on earmarks and lack of a competitive grants program Moffat (1992). As observed in a *Washington Post* editorial, (Washington Post 8/4/92), however, "Pork is the epithet you apply to the stuff you don't want to buy and the other fellow does. The stuff you want to buy is sirloin steak."

My view of agricultural research funding is at once simpler and more subtle. I suspect that the words for these decisionmaking mechanisms -- pork, peer review, and priorities -- don't convey the purposes, timing, or processes involved. They fail to account for the diverse goals of the system. They also betray an ignorance, and some disdain, for the diversity of participants in the political system. As my client is quick to remind, public money defines the public's interest. Researchers are just one subpopulation, and an admittedly self-interested one at that.

Just as members of Congress -- through their votes and influence on committees -- claim to represent the public interest, researchers -- through their expert advice to Federal agencies -- claim to represent the best interests of their research communities. Both categories of guardian overstate their representativeness, the correctness of their methods, and the wisdom of their advice. In short, the funding mechanisms advocated and employed, while defensible, are flawed. And a mechanism suited to one purpose may be ill-suited to another. A corollary is that what works in one agency's culture may not work effectively in another's. Although we may each admit this, we nonetheless all have our preferences and reasons for advocating one mechanism over others, see e.g. Bozeman & Crow, and Mervis (1992).

Peer Review and Alternative Decision Tools

If we step back from funding per se, we can see an array of mechanisms for allocating Federal research support. Each is an alternative to the use of expert judgments solicited from outside an agency to inform project selection, what is commonly called "peer review." These alternatives are:

A. Formula Funding -- formal, non-merit based review

B. Earmarking -- formal, non-peer review

C. "Old Boys" Network -- informal, peer-based non-review

D. Manager's Discretion -- informal in-house review with unknown criteria and participants

To elaborate, formula funding, legislated through the Hatch (1887), Smith-Lever (1914), and McIntire-Stennis (1962) Acts, uses State population size to determine the amounts of USDA money awarded to its land-grant universities. Technical merit is not a consideration; the application of research funding to local farm problems is.

Earmarking is a procedure by which Congress designates funding directly in an appropriations bill for a facility or a research project. This is formal, non-peer review because expert peers are not consulted about these allocation decisions, which are almost always motivated by concerns for a local institution (declared, of course, in "the national interest.") Earmarks are usually imposed on Federal agencies (discussed below.)

A panel of expert peers can nevertheless form an "old boys" network, which is tantamount to informal, peer non-review that weighs heavily criteria other than technical merit of the proposed research in funding recommendations. These other criteria center on characteristics that relate mostly to the past performance ("track record") and current institutional affiliation of the applicant, rather than merits of the proposed research. This is a familiar criticism of national, centralized peer review systems, as practiced by the National Science Foundation (NSF) and the National Institutes of Health (NIH).

Finally, manager discretion is a euphemism for informal in-house review since the criteria and the participants involved in the decisionmaking are unknown to outsiders. It is assumed that program managers with well-defined goals must be more responsive to the general agency mission and particular program objectives than to outside peer evaluations. The Defense Advanced Research Projects Agency (DARPA), which supports research on emerging and risky technologies for the Department of Defense (DOD), may be the best case in point.

Because all of these mechanisms shape the Federal research portfolio, it is naive to embrace or discount any single one. While peer review is not infallible as a decisionmaking tool, a peer-based system is often touted as the only effective, efficient, accountable, responsive, robust, rational, fair, valid, and reliable guarantor of quality in science. Any peer review system would strike a compromise among these properties. For instance, accountability demands some sacrifice of efficiency, robustness reduces responsiveness, and fairness may cost some effectiveness.

Consider, too, the variations in purpose and practice that might characterize a peer review system. For example, are the technical reviews binding, i.e., tantamount to "approval" or "veto power," or merely advisory, i.e., augmented by other criteria? Is a single program manager ultimately responsible for decision or is responsibility entrusted to a chartered, i.e., legally-constituted and government-approved, committee or to an ad hoc panel? Are reviews done anonymously as well as confidentially, with opportunity for revision, rebuttal, and/or appeal, or does the proposer know the identity of the peers and could communicate directly to them about the proposal? Is the pool of reviewers limited to those located in the nation awarding the funds, or does it include international peers? How an agency approaches such questions will doubtless influence its funding decisions. Indeed, there are several outstanding issues in the practice of peer review that beg for empirical examination. Conflicts of interest are created by the multiple roles that people play as participants in peer review systems. This is really a question of ethics or what is proper conduct, which is defined by and therefore relative to a research community, Chubin (1985). The definition of who is a peer is problematic since substantive expertise must now be supplemented by other (e.g., institutional, regional, and demographic) characteristics in the aggregate composition of review panels (or even mail reviewers). Part of coping with knowledge specialization, fragmentation, and interdisciplinarity also means asking: Peer review for what? What are the objectives of that particular peer review system? Brooks distinguishes between reviewing for "truth" v. reviewing for the "utility" of research, Brooks (1978). And how do the criteria for selection change given the objectives that the agency or program wants to pursue?

Management issues include the use of reviews and decision rules to affect such things as award durations, amounts, research costs, and accountability for funds. Agency traditions, structure, and discretion all contribute to definitions of "best practice," Atkinson and Blanpied (1985). Can we think of risk-taking and experimentation as creative manipulation of peer review systems? How do the preparation, credentials, and power of agency research officers influence these systems?

Evaluation has a role to play in connecting *ex post* evaluation (i.e., after the research is completed) to inform *ex ante* or peer review decisions (i.e., judgments of technical merit, investigator capability, and research "promise"). But how do resource constraints affect the process and interpretation of peer reviews? For example, see NSF (1988 and 1990). How much knowledge of agency/program budgets should reviewers have? And is the Federal role to support the "cutting edge" or to "bootstrap" unfashionable basic research? These are philosophical as well as pragmatic questions. Different answers reflect the pluralism of the Federal research system and the unique role of each agency within it, see OTA (1991).

Peer review is clearly not the pristine, omniscient system that its adherents claim it to be. For some purposes, it is the obvious quality control mechanism and better than the alternatives. For other purposes, it is neither sufficiently discriminating nor functional for the achievement of certain research-based outcomes. We must also recognize that there are limits to fine-tuning. Reforms of existing systems, anchored in agency cultures, occur "at the margin." In contrast, what would an agency peer review system designed from scratch look like? How would its funding criteria reflect short-term priorities and longer-term, i.e., mission, goals? And how would it differ from the models with which we are all familiar and are sometimes frustrated by? Remember, too, that at stake is a process that should be readily understood, if not altogether transparent, to its participants. Unfortunately, there is evidence that questions this (see McCullough). Win or lose -- most researchers experience both in applying for Federal research support -- peer review must sustain participation. Faith in the process cannot hinge on outcomes alone.

Funding Agricultural Research

The majority of USDA's Cooperative State Research Service (CSRS) support takes the form of formula funds directly appropriated by specific acts of Congress. Special Research Grants consist mostly of line item appropriations, which also require oversight by CSRS. The Competitive Research Grants Office conducts nationwide competition for basic research funds in specific fields. The National Agriculture Research Initiative would enlarge the Grants Program, but the pace has been slower than prescribed by the NRC, (Abelson 1992).

Contrast formula funding with "set-aside" programs. Set-asides define the eligible pool of competitors according to some characteristic of underparticipation, e.g., gender, race/ethnicity, age, type of institution, or geographical region. Such programs have proliferated at NSF and NIH, the two agencies that practice nationally competitive, peer-based decisionmaking. The philosophy underlying set-asides and formula funding, as noted in Federally Funded Research, is this:

> *Both set-asides and formula funding represent a form of legislated and/or within-agency recognition that certain research goals cannot be achieved via conventional proposal review. Thus, agency programs are created to direct funding that satisfies longstanding or emerging needs in novel ways. Such departures are almost always seen as diluting quality, i.e., trading off excellence in research for the fulfillment of 'subsidiary' agency objectives. But at what point do these objectives become central to the agency mission, or address multiple deficiencies in the distribution of research funds and the execution of research? OTA (1991, pp.130-132).*

Finally, **earmarking** is thrust upon funding agencies by the political system. Many in the scientific community see this practice -- myopically -- as circumventing the quality control signified by peer review, see Chubin and Hackett. The distribution of academic earmarks by appropriations subcommittee for the 1980s decade **(Table 1)** shows that one-quarter of all such funding and one-half of all the bills containing academic earmarks originated with the Agriculture Subcommittee. Trends in Federal R&D obligations over a 30-year period allow some comparison of USDA funding to that of other major research agencies **(Figure 1)**. Accounting for inflation, USDA's budget has been flat.

A summary of two funding distributions -- State recipients of Federal R&D funds and of academic earmarks **(Table 2)** -- illustrates the phenomenon of what Cong. George Brown, chairman of the House Science, Space, and Technology Committee, calls "double dipping." States with historical strength as university research performers also pursue earmarks to refurbish or expand its infrastructure. The result is that earmarks directed to "have-not" States do not redress inequities in the allocation of research funds. Congressional sensibilities on pork, however, continue to grow, Science 10/16/92, Pianin, and Cordes.

These, of course, are all input data. On the output side, OTA (1992) states:

> *It is widely presumed that the research supported via a competitive grant mechanism is of higher quality than that funded by formula funds . . . However, it is also possible that competitive grants distort the research mix favoring disciplinary research over problem-solving research. . . .*
>
> *. . . recent research completed by OTA and the University of Minnesota suggest that the most appropriate policy is a mixture of formula and competitive grants, with different funding mechanisms potentially more appropriate for different functions and goals of land-grant universities. . . .* **[see Table 3]**
>
> *If the goal is to increase cutting-edge research, competitive grants might best be emphasized. If the primary goal is to enhance research applicable to problem solving (more development and adaptive research and technology transfer) or to train future researchers, the more stable and locally controlled Hatch funds may be the more appropriate mechanism. . . . [T]he two types of grants depend on the priority system given to the multiple missions of the experiment stations.*

So once again, if we step back from agricultural research and look across all the R&D agencies, principles of decentralization, pluralism, and participation -- local adaptations and responses to national missions -- characterize all Federal research funding. Set-asides such as NSF's Experimental Program to Stimulate Competitive Research (EPSCoR), increase research capacity in traditionally have-not States originated at NSF in 1978. More than a decade later, Congress began mandating EPSCoR at the other R&D agencies, OTA (1991). Similar programs established by the imminent NIH and NSF strategic plans and the Commission on the Future of NSF were responses to demands for managing scarce resources while making research more relevant to industry, linking research to technological innovation and its transfer, and enhancing U.S. economic competitiveness internationally. All signaled some restiveness and rethinking of agency missions in terms of national needs. This change and uncertainty will make researchers nervous. Should it make agricultural researchers opt for one funding mechanism over another? I hope not. Instead, they should embrace priority-setting as appropriate planning for a lean foreseeable future. Shifts toward applications and utility in the distribution of resources within the Federal research portfolio do not spell the undoing of agency missions or the abandonment of basic research. Rather, such shifts represent changes in what and how the Federal Government chooses to support.

Conclusions

From the foregoing, the national trend toward the form of research performance that has been the centerpiece of the agricultural research system should be clear: collaborative, local, consumer- and client-oriented. Indeed, OTA has called attention to the growth in an "industrial model" of research -- center-based and often interdisciplinary, OTA (1991). The State Agricultural Experiment Station is one prominent source of this model. As one observer has put it:

Our agricultural revolution was one of America's most successful R&D programs... The Agricultural Extension Service was a group of professionals dedicated to working one-on-one with farmers in their regions. . . These peripatetic professionals were the glue that held together the land-grand universities, the government agencies and the farming community in their cooperative enterprise. Above all, they spoke the languages of both the laboratory and the field and thus helped the two cultures to work with each other, Michaelis (1992).

To conclude, within the Federal portfolio of agricultural research, clarification of programs, purposes, and their administration will improve as priorities -- and the discretionary budget -- drive what USDA and its industrial partners seek to accomplish. Questions of relevance, contribution to education and human resources, as considered by Clemmit and Veggeberg, efficiency, and responsiveness must be explicitly weighed along with technical merit as criteria in funding decisions. Progress in agricultural research and technologies, especially biotechnology, fuels such differentiated decisionmaking. A diverse Federal research portfolio demands it.

If the notion of a research portfolio is sound, then we cannot forget the range of the nation's research goals and different ways of fulfilling them. There is no single "best practice": formula funding, peer review, and earmarking are not panaceas. They each will continue to have a place in the support of the diverse agricultural research system.

Figure 1

Federally Funded Research in the Major Research Agencies: Fiscal Years 1960-90 (In billions of 1982 dollars)

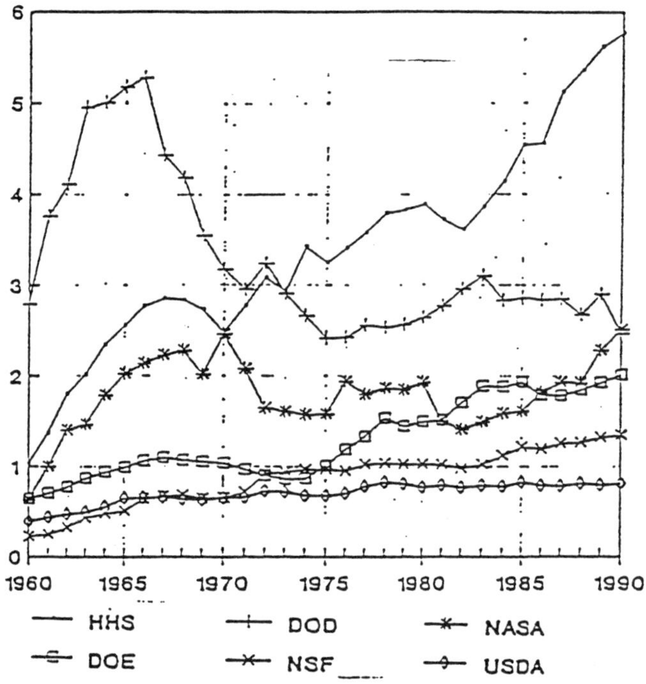

KEY: HHS-U.S. Department of Health and Human Services; DOD-U.S. Department of Defense; NASA-National Aeronautics and Space Administration; DOE-U.S. Department of Energy; NSF-National Science Foundation; USDA-U.S. Department of Agriculture.

NOTE: Research includes both basic and applied. Figures were converted to constant 1982 dollars using the GNP Implicit Price Deflator. 1990 figures are estimates.

SOURCE: National Science Foundation, *Federal Funds for Research and Development, Detailed Historical Tables: Fiscal Years 1955-1990* (Washington, DC: 1990), table A; and National Science Foundation, *Selected Data on Federal Funds for Research and Development; Fiscal Years 1989, 1990 and 1991* (Washington, DC: December 1990), tables 4 and 5.

PUBLISHED IN: U.S. Congress, Office of Technology Assessment, Federally Funded Research: Decisions for a Decade (Washington, DC; USGPO, May 1991), fig. 1-3.

Table 1

Apparent Academic Earmarks, by Appropriations Subcommittee: Fiscal Years 1980-90 (number of projects) dollars, in millions)

Fiscal year	Total (all bills)	Energy and Water	Agriculture	Defense	Interior	Labor-HHS-ED	Treasury-Postal	VA-HUD Transportation	Independent Agencies	Commerce
1980	(7) $10.7	-	(5) $4.2	-	-	(2) $6.5	-	-	-	-
1981	-[a]	-	-	-	-	-	-	-	-	-
1982	(9) 9.4	(1) 1.1	(7) 7.3	-	(1) 1.0	-	-	-	-	-
1983	(13) 77.4	(1) 6.1	(3) 11.6	-	(4) 26.1	(5) 33.7	-	-	-	-
1984	(6) 39.3	(2) 10.0	(1) 1.0	(1) 0.8	-	(1) 19.0	0	(1) 8.5	-	-
1985	(39) 104.1	(9) 29.6	(16) 39.0	-	(7) 17.7	(4) 5.6	-	(1) 0.3	(1) 8.0	(1) 4.0
1986	(39) 115.9	(6) 35.3	(16) 15.5	(9) 39.8	(6) 18.5	(3) 8.8	-	-	-	-
1987	(50) 170.4	(9) 58.7	(21) 38.0	(9) 56.6	(5) 12.5	(3) 4.0	-	-	-	(3) 0.6
1988	(72) 232.3	(28) 128.3	(28) 49.2	(2) 28.5	(6) 6.3	-	(1) 4.0	(3) 8.6	(1) 1.7	(3) 5.6
1989	(87) 202.5	(9) 73.9	(49) 51.9	(3) 21.5	(6) 9.9	(2) 4.8	(9) 38.6	-	-	(9) 2.2
1990	(84) 132.4	(7) 37.5	(60) 47.5	-	(9) 6.3	-	(7) 23.1	(2) 5.0	(5) 10.2	(4) 2.8
Total	(416) $1,094.4	(70) $380.4	(206) $265.2	(24) $147.2	(44) $96.3	(20) $82.1	(17) $67.7	(7) $22.4	(7) $19.9	(20) $15.3
Percent of total .	100%	34.7%	24.2%	13.4%	8.8%	7.5%	6.2%	2.0%	1.8%	1.4%

[a]The only direct appropriations in 1981 were to historically Black universities (three) and two other institutions within the District of Columbia, which Savage does not count as academic earmarks.

KEY: ED-U.S. Department of Education; HHS-U.S. Department of Health and Human Services; HUD-U.S. Department of Housing and Urban Development; VA-U.S. Department of Veterans Affairs.

SOURCE: James Savage, Office of the President, University of California, "The Distribution of Academic Earmarks in the Federal Government's Appropriations Bills, Fiscal Years 1980-1990," mimeo, Mar. 7, 1992, table 2.

PUBLISHED IN:
U.S. Congress, Office of Technology Assessment, Federally Funded Research: Decisions for a Decade (Washington, DC: U.S. Government Printing Office, May 1991), table 3-3, p. 91.

Table 2

**Distribution of Federal R&D Funds (FY 1987)
and Earmarked Funds (FY 1980-89), by State**

	Rank in Federal R&D Funds Received FY87	Rank in Academic Earmarks Received FY80-89
California	1	16
New York	2	2
Maryland	3	32
Mass.	4	1
Texas	5	22
Penn.	6	11
Illinois	7	5
Michigan	8	36
N. Carol.	9	25
Ohio	10	29

Notes

a. Top 10 States Ranked by Federal R&D received in FY87 also accounted for 35 percent of academic earmarks. States ranked 11-20 accounted for 11 percent, 21-30 received 28 percent, and remaining 21 States (incl. District of Columbia shared 25 percent).

b. Top 10 States Ranked by Academic Earmarks Received in FY80-89 accounted for 63 percent of all earmarks. Other States in Top 10 are: Oregon, Florida, Louisiana, S. Carolina, W. Virginia, Alabama, and New Hampshire. Five States received no academic earmarks during the decade. States ranked 11-20 accounted for 23 percent, so remaining 26 States shared 14 percent of the earmarks received. Total cost for the decade: $905 million.

Source: U.S. Congress, Office of Technology Assessment, Federally Funded Research: Decisions for a Decade (Washington, DC: U.S. Government Printing Office, May 1991), table 3-4.

Table 3

Mean Values of Selected SAES Output by Grant Type

	Hatch	Competitive	Other
Citations per article[a] ..	1.70	3.98[e]	1.82
Articles per grant	2.47	2.14	2.24
Weighted articles per grant[b]	4.83	8.33[f]	4.74
Journal publications per grant[c]	4.70	4.52	3.68
Weighted publications per grant[d]	7.07	10.62[g]	6.58
Degrees per grant	0.45[e]	0.18	0.25
Bulletins per grant	0.35	0.09[e]	0.28

[a] Articles are articles published in peer reviewed journals.
[b] Weighted articles are published articles weighted by citations.
[c] Journal pubs are published articles, articles submitted, articles in press, and abstracts in peer reviewed journals.
[d] Weighted pubs are articles submitted, articles in press, and abstracts in peer reviewed journals, and published articles weighted by citations.
[e] Significantly different from other two groups at 95% confidence level.
[f] Significantly different from other two groups at 94% confidence level.
[g] Significantly different from other two groups at 92% confidence level.

SOURCE: Marie Walsh, "Factors Affecting the Cost and Productivity of Biotechnology Research at the State Agricultural Experiment Stations", PhD thesis, University of Minnesota, in progress.

PUBLISHED IN: U.S. Congress, Office of Technology Assessment, A New Technological Era for American Agriculture (Washington, DC: USGPO, August 1992), table 16-14.

References

Ableson, Philip H. 1992. "Agricultural Research." *Science*, vol. 257, Aug. 28, 1992, p. 1187.

Atkinson, Richard C. and William A. Blanpied. (Summer) 1985. "Peer Review and the Public Interest." *Issues in Science and Technology*, vol. 2, Summer 1985, pp. 101-114.

Bozeman, Barry and Michael Crow. (Fall) 1992. "Science Policy: Pork Barrel or Peer Review." *Forum for Applied Research and Public Policy*, vol. 7.

Brooks, Harvey. 1978. "The Problem of Research Priorities." *Daedalus*, Vol. 107, pp. 171-190.

Clemmit, Marcia. (September) 1992. "Plant Science Job Horizon Dimmed by Lack of Funding." *The Scientist*, Sept. 14, 1992, pp. 1, 6-7.

Cordes, Colleen. "Earmarks Seen Helping Colleges To Hold Down Overhead Rates." The Chronicle of Higher Education, p. A23.

Chubin, D. E. and E. J. Hackett. 1990. *Peerless Science: Peer Review and U.S. Science Policy.* Albany, NY: SUNY Press.

Chubin, D. E. 1985. Research Malpractice." *BioScience*, vol. 35, February 1985, pp. 80-89.

Fox, G. et al. (June) 1992. "Balancing Basic and Applied Science." *BioScience*, vol. 37, 1987, pp. 507-509.

Hadwiger, Don F. 1992. "Technology in a Fragmented Politics: The Case of Agricultural Research." *Technology in Society*, vol. 14, 1992, pp. 283-297.

Holt, Don A. "Achieving Focus in Agricultural Research and Extension." Presented at the annual meeting, Great Plains Agricultural Council.

Mervis, Jeffrey. (November) 1992. "Funding of Two Science Labs Revives Pork Barrel Vs. Peer Review Debate." *The Scientist*, pp. 1, 4-5.

Moffat, Anne Simon. (June) 1992. "New Plant Institute Recommended." *Science*, vol. 256, pp. 1755-1756.

McCullough, J. 1989. "First Comprehensive Survey of NSF Applicants Focuses on Their Concerns About Proposal Review." *Science, Technology, & Human Values*, vol. 14, pp. 78-88.

Michaelis, M. 1992. "R&D To Stay Competitive" (letter). *Washington Post*, 4, p. A24.

Office of Technology Assessment. "Proposal Pressure in the 1980s: An Indicator of Stress on the Federal Research System." Staff paper, April 1990).

Pianin, Eric, et al. (June) 1992. "Academic Pork Gets Its Own Peer Review." *Science*, vol. 258, Oct. 16, 1992, p. 387.

Pianin, Eric. (September) 1992. "'Academic Pork' Fills Favored School Larders." *Washington Post*, p. A17.

Task Force. (August) 1990. Proposal Review at NSF: Perceptions of Principal Investigators (Washington, DC: NSF 88-4, 1988)

Task Force. (August) 1990. Report of the Merit Review Task Force (Washington, DC: NSF 90-113).

Veggeberg, Scott. 1992. "Plant Science in Need of Healthier Funding Climate." *The Scientist*, pp. 14, 18.

U.S. Congress, Office of Technology Assessment. (May) 1991. *Federally Funded Research: Decisions for a Decade* (Washington, DC: U.S. Government Printing Office).

U.S. Congress, Office of Technology Assessment. (August) 1992. A New Technological Era for American Agriculture (Washington, DC: U.S. Government Printing Office, p. 422.

"R&D To Stay Competitive." *Washington Post*, Aug. 16, 1992, p. C6.

"The Pork Rind Question." *Washington Post*, Aug. 4, 1992, p. A18.

Chapter 18

The National Research Initiative: Competing at the Margin

Susan Offutt
Executive Director, Board on Agriculture
National Research Council

In no other sector of the American economy has there been a longer or more beneficial relation than that between agriculture and public research institutions in land grant universities and Federal agencies. Widely credited as the engine that has driven the substantial achievements of American farmers, the nation's agricultural research system has, with growing frequency, been cited as a model in the design of similar research and development partnerships for other parts of the industrial base. However, it diminishes none of the achievements of the past to suggest that agriculture should now concentrate on its future.

The part of the future of concern here is the mechanism by which agricultural research is funded. Sustained over a century by Federal formula grants matched by state contributions, the system has more recently experimented with competitive grants. In 1989, the Board on Agriculture of the National Research Council made a proposal to mobilize the nation's scientific and engineering communities to advance the quality of agriculture, the food supply, and the environment through the significant expansion of competitive research grants.[1] Intended to garner new monies for agricultural research, this National Research Initiative (NRI) was to use $500 million annually, distributed among six fields of endeavor: plant systems; animal systems; nutrition, food quality, and health; natural resources and the environment; engineering, products, and processes; and markets, trade, and policy. Authorization for the full $500 million program was included in the 1990 farm bill.

The Board argued that implementation of the NRI would assure the continued benefits of the high return to investment in agricultural research, encourage the participation of the entire science community in agricultural work, and advance US agriculture while contributing advances in relevant scientific fields, such as biomedicine and ecology. The hope was that, when fully funded, the NRI would make grants of larger size and duration than under the existing competitive grants program within USDA and that those grants would be made to individual principal investigators, to multidisciplinary teams working on basic or mission-linked research, and to institutions and individuals to strengthen the US research capacity.

The Board's proposal was endorsed by the Department of Agriculture, with which the Board had consulted closely in formulating the plan for the NRI. The Busch Administration subsequently proposed to the Congress that the existing competitive research grants program be expanded by $50 million annually. And, the Congress responded with an increase in the appropriation from about $40 million to $75 million. The 1990 farm bill included authorization for the NRI program content and ultimate funding goal of $500 million. In FY1992, the NRI funding level was set at $97.5 million.

Considered against the backdrop of an increasingly constrained Federal budget allocation for all agricultural programs, the NRI has enjoyed remarkable success. By FY1993, however, the strictures of the 1990 budget summit were being felt: NRI funding stalled at the previous year's level of $97.5 million. While all six categories received some measure of funding by FY1992, the NRI was still far from its overall goal and from being able to fulfill the Board's hope that individual grants would average $100,000 per year (compared to the current average $50,000) and last longer (for three to five years, compared to the current average one to two years). Apparently poised for genuine success, the NRI faces the certainty of continued constrained Federal budgets and perhaps, then, the prospect of stagnation and failure to fulfill its considerable promise.

Prospects for the NRI

In the absence of real growth in the Federal budget, programs must compete against each other for additional dollars. Although the Board had called for new funds for the NRI and had suggested transfer of expenditures from commodity subsidies to research, such resources are not forthcoming. Assessing the prospects for the NRI, then, requires an analysis of the existing state of Federal funding for agricultural research. Although the concentration here is on Federal funding, it is clear that state funds are in many, if not most, cases the most significant source of support for land grant research. However, the Federal situation is most important to the NRI and to a reading of the relative merits and support for alternative means of funding agricultural research.

The Bush Administration's original support for $50 million annual increases in the NRI was conditioned on the appropriation's being free of earmarks. Congressional earmarks, also known as "pork barrel" projects, have been aptly defined as "spending with a zip code." These earmarks appear in appropriations bills, or more usually in the reports that accompany them, and give the executive branch agencies specific directions on how and where Federal funds are to be spent. Although report language does not carry the weight of law, realpolitik dictates the agencies follow the instructions as if they were indeed binding. Nevertheless, the Bush Administration's condition is grounded in the belief that earmarking runs exactly counter to the intent of competitive grants, with general priorities ordered to address the public good and allocation within categories based on peer review to identify the best science. Of course, the Congress views earmarks as a legitimate means of achieving those same goals, and it is this difference of opinion that sets up the competition for Federal funds, at the margin, between competitive grants and earmarks.

The nature of this competition is the subject of this analysis. A comparison of the substance of the earmarks to the program content of the NRI is intended to shed light on differences (or similarities) in research priorities and research grant size and recipients. Only with this kind of information can an assessment be made of the opportunity costs of funding research through earmarks rather than competitive grants. There is no argument that investment in individual earmarked projects quite likely yields a positive return. Rather, the question is whether an even larger return could be had by putting the same dollars into competitive grants. With constrained budgets, total benefits can only be increased through reallocation. Not to give away the punch line, but this analysis will be suggestive, not conclusive. The hope is that such an exercise leads to a serious consideration of how earmarking should be interpreted as an expression of priorities for research and for institutional structure.

Agricultural Research Earmarks

Congressional earmarking of research grants is not confined to agriculture by any means. However, it has historically found its fullest expression there. An analysis by the Office of Science and Technology Policy found that FY1992 total science earmarks for both facilities and research amounted to $993 million, of which the single largest portion (almost 20 percent) went for 334 agricultural projects at a cost of $180 million.[2] And, within the appropriation for agriculture, earmarks are not confined to the Cooperative State Research Service (CSRS), the primary source of funds for the land grant universities, but also are found in the Agricultural Research Service (ARS) and the Extension Service (ES). Indeed, the FY1993 ARS appropriations report contained 61 separate instructions on spending, of which 49 identified specific locations or recipients.[3]

The CSRS appropriation ($485 million in FY1993) accounts for about one-third of Federal spending on agricultural research (CSRS, ARS, and ES combined). Of the $485 million, $52 million is allocated to construction and renovation of (primarily) university buildings and facilities. Except for the one time in FY1991 the Bush Administration proposed a competitive building grants program (rejected by the Congress), it has not requested any support for university construction. The Congress has set its own priorities and given its own instructions, requiring (through non-binding report language) matching of Federal funds and the complementarity of the facility's program with that of USDA. Given the documented need for attention to the nation's agricultural research facilities, a full analysis of these construction earmarks would be instructive, but is not pursued here. However, there is obviously a close relationship between the state of facilities and the ability of the system to conduct first-rate research under the NRI, a link which led to the Administration's proposal for a companion competitive construction grant program.

Of the $433 million in program funds for CSRS, formula grants under the Hatch Act account for $168 million, a total little changed, in real terms, over the past decade. Of the remaining $265 million, $97.5 million goes to the NRI and about $64 million is earmarked for special grants "with a zip code." These earmarks, some 120 in number, thus account for about 15 percent of the CSRS program spending.[4] As a matter of methodology, it should be noted that identifying "pork" is a somewhat subjective matter, but the judgments applied here are consistent with most analyses.

Comparing Earmarks and the NRI

Size and Duration of Grants

CSRS earmarks average about $460,000 each, although if the single largest grant (at $4.2 million for wood utilization research) is not counted, then the average falls to $306,000. The smallest grant, of $25,000, went to Virginia for procerum root disease research. Length of grants are not specified, although most are effectively one year in duration, as the preponderance of these earmarks appear in appropriations reports year after year. In comparison, the NRI grants (based on data for FY1991) last for an average of two years at an annual average amount of $53,000. Special grant recipients, then, receive more than could be expected from successful competition in the NRI and can reasonably expect funding to continue over several years once the claim has been established.

Research Subject

To compare research priorities, each CSRS special grant was assigned to one of the six NRI program areas, based on the title and the information in the report on special grants provided each year to the Congress by CSRS. Accounting for the difference in total dollars between the special grants and the NRI, the proportion of resources devoted to each category was calculated, as given below.

NRI Program Area	NRI Percent	NRI Rank	Special Grants Percent	Special Grants Rank
Plant systems	41	1	42	1
Animal systems	26	2	15	3
Natural resources	18	3	11	4
Nutrition	7	4	3	6
Markets	4	5	19	2
Processes	4	6	10	5

While both the NRI and special grants allocate the largest portion of resources to plant systems, the relative importance of each of the other five program areas differs between the two. Most notably, the markets, trade, and policy area, supporting mostly social science research, does relatively better under special grants than the NRI. Nutrition, in contrast, fares more poorly in special grants, as is the case for natural resources. One way of interpreting these results is as a commentary on the relative political skills across agricultural science disciplines. Another is as a measure of the relative strengths of the likely beneficiaries of research, with the beneficiaries of nutrition and natural resource work more numerous and less well-organized than the commodity groups that are effective proponents of plant and animal research.

Geographical Distribution

In contrast to the outcome with respect to grant size and subject matter, the geographic distribution of grants between the NRI and CSRS earmarks are quite comparable. Using USDA regional categories, the NRI reported the following distribution across land grant universities (70 percent of the total) for the FY 1991 program: North Central (36 percent of all grants); South (29 percent); West (23 percent); and Northeast (12 percent). The same ranking and shares are seen in the FY1993 CSRS earmarks, which go largely, although not exclusively, to land grants.

These rankings can be taken to reflect the existing distribution of agricultural research capacity across the country. However, more detailed information on the identity of individual recipient institutions is required to determine whether intra-regional re-allocation is taking place between those who win NRI grants and those who do not. The Board on Agriculture recognized that not all institutions might compete equally well and, to address this point, included a category of strengthening grants in its NRI proposal. And, of course, the distribution of special grants also mirrors the composition of the appropriations sub-committees and the home states of the Congressional leadership.

Implications

NRI and earmarked special grants exhibit differences that are potentially significant in considering the opportunity costs of not funding agricultural research through competitive grants. While acknowledging the important role that formula funding plays in maintaining the nation's agricultural research capacity, the Board clearly argued that, at the margin, the nation's additional dollars should be invested in competitively awarded, peer-reviewed research. "Spending with a zip code" does not figure into this prescription for maximizing the national gains from investment in agricultural research.

Nonetheless, the message that earmarking sends about where research should be conducted and on what topics is worth considering carefully. The hope is that the cursory analysis presented here stimulates interest in evaluating the relative returns across fields in agricultural science, in terms of their size and the identity of the beneficiaries. To date, most analyses of the returns from investment aggregate research expenditures, yet program and policy officials require guidance on allocation across fields. And, these traditional analyses have generally assumed consumers as the beneficiaries of research that lowers food prices, whereas many of today's research problems address issues of food quality and of environmental problems. How robust is the assumption about the distribution of gains?

With respect to tactical aspects of funding agricultural research, the differences also tell quite a bit about why the NRI at the size and in the form proposed by the Board on Agriculture is the only effective antidote to earmarking. Clearly, earmark recipients receive larger grants and can reasonably hope for some year-to-year continuity, a goal to which the NRI now only aspires. A more detailed look at the identity of grant recipients (and a check against the location of facilities construction and renovation) could reveal much about the capacity of the agricultural research system (as well as the nature of agricultural politics).

ENDNOTES

1. Board on Agriculture, National Research Council. *Investing in Research*. National Academy Press: Washington, D.C., 1989.

2. Office of Management and Budget. *Budget of the United States Government Fiscal Year 1993*. Washington, D.C.: US Government Printing Office, 1992, pp. 139-141, Part One.

3. As an indication of the importance to the Congress of these earmarks, consider the Senate's instruction that while ARS should follow the directions given it in the House report, if the result were to move spending from one state to another, the agency must deliver prior notification to both houses.

4. US Congress. *Conference Report to Accompany H.R. 5487, Agriculture, Rural Development, Food and Drug Administration, and Related Agencies Appropriation Bill, 1993.*

Chapter 19

Expanding Agricultural Research Agendas With A Shrinking Resource Base: Current Reality for the Public Sector

Katherine Reichelderfer
Economic Research Service
U.S. Department of Agriculture

The driving forces behind recent and anticipated change in the demand for the products of agricultural research are covered in other chapters in this volume, (Busch, Woods, Duvick). In essence, we can expect a predominantly urban/suburban public, to whom historical preoccupations with food cost, food availability, and the subsequent desire for rapid agricultural productivity are foreign, and for whom contemporary concerns with the environmental, human health, and social consequences of agriculture are paramount, to be increasingly well represented in Federal and State political systems. At the same time, the capacity for supplying public agricultural research is being and will increasingly be pressured by public budget crises and the greater relative success of health care and social security constituencies in competing for a declining pool of public funds (Reichelderfer). The implication of these simultaneous trends is that the public agricultural research community must find some way to meet increasing demand with constrained resources.

In this paper, I discuss a strategic approach which relies upon using existing or diminishing levels of resources to expand the clientele base served by the system. This resource-constrained approach recognizes that traditional methods of agenda change through expansion of program size are not feasible in the short run, but asserts that reduced capacity need not be equated with reduced capability.

Why Expand the Clientele Base?

This discussion relies on a prosaic but commonly accepted understanding of public agricultural research; that is, research is conducted by and on behalf of "the public." A main justification for expansion of the clientele base for such research is that this "public" has changed substantially more over the past several decades than has the system serving it.

Peter Drucker calls farmers one of society's "ancient classes," since "for millennia, farmers ... *were* (Drucker's emphasis) civilization" (Drucker, p. 188). It has only been in the last 110 years or so that farmers have gone from a majority to a 2-percent minority of the U.S. population. Thus, it has been in relatively recent times that "farmers" and "the public" have lost synonymy.

Those who contend that agriculture has not received its "proper proportionate share" (Woods) of Federal research funds in recent years occupy a supportable position only if "agriculture" is viewed as involving far more than commercial farmers and agribusiness. If "agriculture" is defined by farmer numbers or by the contribution of farm-related business to gross domestic product, then, in fact, it can only be concluded that rising research funds have been going to a contracting sector. The argument of declining "proper proportionate share" holds up only if we use a much broader definition of "agriculture."

A broad concept of agriculture would include not only the producers, processors and marketers of commercial agricultural goods, but also those vastly more numerous consumers of agricultural goods, and those sector-external individuals and groups affected both positively and negatively by the social and environmental consequences of agricultural activities. These latter groups can be called "nontraditional" with respect to a narrowly defined "agriculture." But, as McDowell puts it, "Unless the non-traditional clients can be served by research... and their support organized in coalition with agricultural interests, the system may not be able to make the adjustments necessary to continue to serve even traditional audiences" (McDowell, p. 21).

The bottom line is that unless the public agricultural research agenda becomes appropriately associated with practical problems of a contemporary, largely urban/suburban public, it will become more and more difficult to generate and maintain funding for long-term and basic agricultural research.

Clientele Expansion Not Equivalent to Clientele Replacement

Expanding the clientele base for agricultural research does not mean, and, I believe, should not mean, abandoning traditional agricultural clientele. For one thing, such traditional clients as farmers, farm businesses, agricultural trade associations, and commodity groups retain considerable absolute levels of political clout in many States and locales, even as their relative aggregate clout may be eroding.

Another reason for holding on to traditional clients is that it is really only through an understanding and the addressing of agriculturally-related systems that we participants in the agricultural research community have a strong comparative advantage relative to colleagues in related fields. The more indistinguishable that agricultural engineering or agricultural economics, for example, become from general engineering or economics, the less likely it is that agricultural engineers or agricultural economists will be singled out as special recipients of public support.

The spirit of clientele expansion, rather than clientele replacement, also suggests as counterproductive the claim (see Woods) that environmentalists or other public interest groups are attempting to reduce agricultural funding levels. It is probably more useful, if not more accurate, to see those groups' interventions as a means for increasing research funding for addressing agriculturally-related problems in the broadest possible context of the universe affected by agricultural activity.

This brings us to the question of how the system can adapt to expand clientele without losing capability for unique contributions under anticipated constant or declining levels of resources. While a variety of strategies can and likely will be employed to achieve such a goal, this paper focuses on one that can be implemented at all levels, from individual researcher to department, college, or agency.

Reorienting Research Perspectives

A critical strategic option for the agricultural research enterprise, particularly in periods of budget stress, is to reorient individual, department, and agency agendas to simultaneously meet both narrowly and broadly defined agriculturally-related social goals. That means looking *at* agricultural issues, but from the point of view of a hugely nonagricultural public, and shifting priorities accordingly.

Examples of how this might be accomplished are numerous. For instance, a research team with long-term focus on the management of dairy production systems might find it challenging to newly apply its expertise to the development, evaluation, and transfer of technology on dairy *waste* management systems within the production context its members already know so well. Such research is of no less value, and may be of greater value to dairy farmers than research strictly on production systems. But, more importantly, its output at the same time addresses the concerns of the far broader group affected by dairy waste problems. The corn geneticist could profitably refocus his/her research away from breeding for increased production, towards breeding for decreased nitrogen requirements, using the same expertise and interests inherent in former research efforts; or hog production system analysts might incorporate animal welfare concerns into their design of cost-effective management strategies. The resource economist whose research has focused on the valuation of recreational use of water, might profitably reorient to include estimation of the extent to which agricultural activities enhance or degrade that value.

This simple proposed strategy relies upon existing interests and expertise. The idea is for agricultural scientists to do what they already do best, but to do that within the context of *both* traditional and nontraditional groups' concerns.

Incentives for and Implementation of Reorientation: A Challenge

At least two main ingredients are necessary to the pursuit of research agenda reorientation to meet multiple client needs. First, one must have a fact-based idea of what each potential clientele group needs or wants as products of public agricultural research. Then, some incentive must exist for individuals and their collective public agricultural research groups to follow up on clientele expansion.

The strategy requires the collection of input from all target or potential clientele groups. It is a demand-driven approach, which means we need to know what all groups want. This knowledge can be achieved in several ways. Focus groups of representatives of similar points of view may be organized and queried about their perceptions and preferences regarding agriculturally-related research. A less expensive option is to use a survey questionnaire to collect information from a broader set of individuals who would otherwise be potential focus group participants. In either case, important goals are to (a) assure extensive diversity of perspective among the individuals and groups from which input is being collected; and (b) build in a mechanism that allows group membership to be associated with attributes of research preferences. In other words, one wants to end up knowing not just the extent to which views about the worth and best direction of agricultural research vary, but also the nature of that variation as it may be associated with specific clientele groups. Groups organized or group membership identified through survey efforts could include such perspectives as representatives of (for example): Farm groups; commodity groups and trade associations; food marketing and processing industries; groups concerned with natural resource conservation/environmental quality; groups concerned with food safety; groups concerned with food assistance/food distribution/hunger; groups concerned with the well-being of rural communities; animal welfare groups; voter organizations; and relevant political decision making bodies.

The process of collecting input for research agenda reorientation does not require that any coalition be built among disparate groups. But it is necessary for the initiator(s) of the investigation to determine the nature of commonality among disparate groups' preferences. This is because it is possible to meet more groups' demands with fewer resources only if the areas of overlap among group preferences are targeted for attention. For example, survey or focus group efforts may determine that livestock producers desire objective identification and analysis of the disadvantages they feel they would face if proposed animal welfare legislation was passed, and that animal rights advocates desire scientific evidence of suspected physiological implications of certain livestock containment systems. A new agricultural research program aimed towards designing and evaluating by cost, yield, physiological, and other measures effective and efficient containment systems could easily meet both groups' preference without in any way requiring that the groups agree with each other or that the research program favor one group's position over the other's. Viewed this way, there is every reason to suspect a high degree of overlap of subject-matter interest by clientele groups can be identified for research agenda reorientation purposes.

Conclusions

Expansion of the agricultural research agenda through reorientation favors and provides some new incentives for interdisciplinary research, relies on maintaining high scholarly credibility, and excludes neither the use of other approaches to increasing research relevancy nor the use of other approaches to research fund enhancement (Reichelderfer). Over the long run, this strategy should help to stabilize resource loss or, ideally, lead to resource gains as the number and size of groups supporting a broadened definition of agricultural research increases.

Making agricultural research agendas more clientele demand-driven also has political advantages upon which the Federal-State Land Grant system could profitably capitalize. There appears to be, in this era of tightening public budgets, a growing belief that "social value" and utility to public interests ought to take precedence over the more traditional criteria of "scientific excellence" and peer-judged merit as bases for public research fund allocation. This shift in the paradigm under which at least Federal research fund decision makers may be viewing the worthiness of alternative research directions is quite clear in a recent report on the "health of research" to the U.S. House of Representatives' Committee on Science, Space, and Technology (U.S. House of Representatives, 1992). This report suggests that "research policy designed forty years ago may no longer be suitable for addressing the problems of today's world," concludes that "new strategies (for addressing increasingly urgent policy dilemmas and societal challenges) should be rooted in the development of explicit linkages between the conduct of federally funded research and achievement of national goals," and recommends establishing a system of research performance assessment that is based on proposed projects' contribution to the public's rather than the scientists' perception of social value (U.S. House of Representatives, 1992).

The agricultural research community is uniquely positioned to take advantage of the client-driven rationale and basis for research funding. It needs only to identify the appropriately broad set of actors with contemporary interests in agriculturally-related issues, and to discover and act upon those actors' perceptions and preferences. That this may be done while utilizing the existing skills and interests of agricultural researchers, and without compromising address of more traditional clients' concerns, adds to the potential for success even in an era of downsizing. It remains completely within the control of the agricultural research enterprise to take the sorts of steps that will maximize the probability of its own success in the future, resource-constrained world.

REFERENCES

Drucker, Peter F. *The New Realities*. 1990. New York: Harper & Row Publ.

McDowell, George R. 1988. "Land Grant Colleges of Agriculture: Renegotiating or Abandoning a Social Contract. *Choices*, second quarter, pp. 18-21.

Reichelderfer, Katherine. 1992. "Agricultural Economics in an Evolving Land Grant System." Keynote address, Annual meeting of the Western Agri. Econ. Assoc., Colorado Springs.

U.S. House of Representatives, 102nd Congress. 1992. "Report of the Task Force on the Health of Research." Chairman's Report to the Committee on Science, Space, and Technology. Wash., D.C.: U.S. Gov't. Printing Office.

Strategies for Collaboration Between the Private and Public Sectors

Chapter 20

Agricultural Research Structures in a Changing World

Brian Wright and David Zilberman
University of California

Public and private sectors in the United States have been major partners in an multibillion dollar agricultural research, development and extension effort that has made possible the impressive rate of technical change seen in the sector over this century. Despite a widely-acclaimed record of high rates of return to public investment (Ruttan 1982), the share of the public partners (USDA and state agricultural experiment stations) fell from 40 to 34 percent over the decade of the eighties (Huffman and Evenson 1993, Table 4.1), due to sporadic cuts in public support levels that are mainly traceable to exogenous budgetary pressures. Over the same period, total expenditures increased, however, from 3.9 to 4.8 billion dollars (at 1984 value) due to an increase in private expenditures.

The budget cuts are the most obvious, but not necessarily the most important forces for change. Other pressures for more fundamental changes in the nature of agricultural research structures are being engendered by qualitative changes in the social demands made on the agricultural research and extension system, in the scope and nature of scientific opportunities for discoveries, and in the potential rewards for researchers.

In this paper, we consider the implications of these changes for the nature of public and private agricultural research. We focus on land grant institutions like U.C. Berkeley and their relations with the private sector in pursuing research and development. The questions include:

- What changes should we want and/or expect in the public and private roles in agricultural research?
- What is the appropriate structure for public-private collaboration in modern agricultural research and extension?
- How should researchers be motivated and rewarded?

In what follows, we first consider the nature of the changes in demands on the agricultural research system. Then we review the reasons why both the public sector and the private sector have valid economic roles in the research and extension system. In Section 3, we consider public and private sector responses to the new research demands and opportunities. Then, focusing on the land grant universities, we consider in turn the implications for institutional structure (Section 4) and for the performance of the universities' roles in teaching, research and extension (Section 5). Conclusions follow.

The Evolution of Research Demands

Historically, the politically expressed demands on the agricultural research system have focused on efficiency in agricultural production and post-harvest activities, and the promotion of rural prosperity and parity with other sectors of the economy.[1] The cost-decreasing and/or yield-increasing innovations the system produced for farmers satisfied the supply requirements so effectively that market forces placed downward pressure on the equilibrium price of farm commodities (Cochrane 1958). This in turn led to effective political action by the farm lobby to obtain protection from price reductions via government intervention to support prices. On the other hand, the distributional concern for relative prosperity of the rural population, featured so prominently in political rhetoric, has not been discernible among the major objectives of public agricultural research, which have remained focused on technical and economic efficiency. Improvements in efficiency clash with the achievement of rural prosperity when the induced increases in supply depress market prices so much as to reduce the net returns to rural suppliers of labor, land, and other inputs.

Now, however, society's effective political demands of the farm sector have become much more complex. Consumers wish to reduce the actual or perceived health risks of chemical residues in foods, and of herbicides and pesticides released into the environment by farmers. Other concerns include the effects of erosion on land and water quality, pollution by animal wastes and, most recently, the effects of methane emitted by belching ruminants on global warming. Standards for animal rights are being advocated for veal calves and poultry. Clearly, the farming process itself is increasingly being subjected to direct social constraints, rather than being viewed as only indirectly socially relevant as the means to achieving a prosperous farm sector and a cheap and secure food supply, (see Busch in this volume).

At the same time, consumers with increasing incomes and no intention of eating or drinking more are looking for higher quality, novelty, variety and constant availability in their foods. These objectives are not obviously mutually consistent, especially if the research continues along the path that produced high yield and low costs.

Scientists and innovators are being asked to furnish production processes, and new products, that respond to these multiple social concerns, and there will be an increasing demand for products and services that can help management in this complex and dynamic environment. Indeed, the multifaceted interactions that constitute an agricultural system will increasingly be the subject of analytical attention. Beyond biotechnology lies the challenge of ecologically appropriate agriculture. This is far more demanding than the more narrow, often organism-specific, focus seen in much of extensive research effort in modern medicine, for example, and also characteristic of that part of agriculture that has in the past achieved the greatest yield increase, the cultivation of a plant or animal species via the exclusion of competitive species by some artificial means. Fortunately these challenges come at a time when the research capacity is being transformed by the revolutions in biotechnology and information processing.

Besides improving the prospects for pursuit of the traditional productivity objectives, the new biological techniques make possible previously unimagined qualitative transformations of plants and animals. They seem to expand greatly the potential for satisfying the new demands for benign production processes, on the one hand, and an array of improved consumption characteristics on the other.

Already scientists have been able, for example, to transfer the pesticidal qualities of Bt into agricultural plants, which might help reduce chemical pesticide use. On the other hand, genetic manipulation has also made possible delivery of better-ripened fruit, such as the Calgene tomato, with less damage and wastage. A slew of more impressive breakthroughs can be anticipated in the years ahead.[2]

Through biotechnology, agriculture will expand to produce new higher-quality forms of existing products, and entirely new agricultural products, many of them beyond the traditional range of agricultural commodities. The term "pharming" refers to the use of plants as factories for biological and chemical products. The biological production of fine chemicals and fibers may offer a less costly alternative to the use of finite or ecologically sensitive environmental resources. Scientists have genetically engineered plants to manufacture a wide variety of materials, including human proteins such as albumin and interferon, alpha-amylase, a bacterial enzyme that is widely used in the food processing industry and natural polymers, including a type of polyester (Moffat 1992).

The advent of the new biotechnological innovations has been fostered by new legal protections in the form of the Plant Variety Protection Certificate (PVPC), established in 1970 and extended in 1980, and subsequently the expansion of patent protection to life forms by the Supreme Court in Diamond v. Chakrabarty in 1980.[3] Similarly the market for software and databases has been sufficiently (if not optimally) developed under the evolving law regarding copyright protection that it has made the personal computer a productive and popular management tool for farmers and farm advisers.

However the legal system also poses new challenges for the modern biological products developed under its protection. The marketing of the Calgene Flavr-Savr tomato may be hindered by legal challenges or the threat thereof, or more generally by adverse publicity regarding the safety of foods containing new genetic material. Already its major backer, Campbell's, is backing away from plans to use this product.

In sum, the agricultural research system is faced with new challenges, but also with an exciting new array of opportunities. How should the agricultural research system be structured in this new environment?

Why Public Research and Extension?

As a preliminary, it is helpful to keep in mind the reasons why we have a public research system at all. After all, we rely on the private sector to produce other products, including food, with profits from private sales as the incentive. Public provision of research and extension has been justified by the argument that the private incentives for research and extension fall short of the public gains at the margin. Important "externalities", benefits (or costs) not captured in private profits, are associated with public research inputs, outputs, or the process itself. This argument has much greater force for some areas of effort than others.

It is widely accepted that pure knowledge, not embodied in any product, is a "public good," the benefits of which are properly made available free of charge because they are "non-rival"; use by one does not reduce the supply available to others. The product of successful basic research is of this type. The desirability of free provision is fortunate, for it is very difficult to exclude non-purchasers from acquiring such "disembodied" information, and since the information is often of quite general use the number of potential "free riders" is often very large. It follows that basic research is mostly produced in, or at least supported by, the public or non-profit sectors.

Basic research findings feed into the applied research areas, which tend to be more industry-specific. In many areas the fruits of applied research are at least partially capturable by its producer, for two quite different reasons. The creator of disembodied, applied process discoveries reaps the benefits to the extent that it dominates the industry that uses the process. For example, an advance in irrigation-equipment manufacturing techniques would be likely to benefit a major manufacturer of such equipment. Furthermore, much applied research and development produces innovations that are embodied in products, such as machines or drugs, that can be sold for profit in private markets, and protected from copying by patents or secrecy. In these cases the derived private demand for applied research may well be adequate, if not optimal.

In agriculture, the producers of applied research have historically had little scope for capturing sufficient compensation from the market to justify their efforts. Most advances have been either yield-increasing or cost-reducing. Some of these advances are embodied in plants or animals that can reproduce, passing on the advances to later users, and spoiling the innovator's prospects for lucrative sales in the absence of effective legal protection. Others are process advances such as new techniques of crop cultivation that can be easily copied by diligent observers. Given the extremely competitive nature of agricultural production, the rewards accruing from use within the innovator's own farming operation are typically a tiny fraction of the full social value.

There are of course prominent exceptions to these generalizations. Private hybrid corn innovators have prospered because their product cannot be successfully reproduced by their customers. New hybrid chicken varieties are also produced privately. Mechanical and especially chemical farm inputs, originating in other sectors, have historically had patent protection. This has not always been very effective. Eli Whitney's (or was it Catherine Greene's? See Warrick 1992 pp. D3-D4) cotton gin, to take a famous example, was so widely copied, despite patent protection, that it was necessary to award him a prize to provide him, ex post, a significant return for his innovation.

Given the anticipated opportunities for innovation, on the one hand, and the lack of privately appropriable returns from many types of applied innovations on the other, the public sector role in supporting agricultural research has been unusually large, and has included the applied development and dissemination of techniques and products that in other sectors is left in the hands of the private sector. Thus the historical role of the public agricultural research complex covers the whole range from basic scientific investigation to application in the farmer's field. In the United States, the land grant universities cover this span, in large part integrated within a college of agriculture and/or natural resources.

The Logic of the Land Grant System

Three aspects of the structure of the land grant agricultural research system suggest the types of externalities important to their mission. The first is that it is a decentralized system of vertically integrated individual institutions, dispersed across the states with substantial funding from state as well as federal sources. Second, its basic structure is program-oriented rather than project-oriented, in that its staffing is predominantly on a permanent basis. Third, the research mission is pursued in concert with an educational mission; researchers are also university teachers and students and others involved in public education.

Decentralization of research and extension to the state level has traditionally been rationalized by the argument that it reflects the fact that many applied research problems are locally specific. Pests, diseases, plant varieties and cultivation practices differ across states and even counties. An institution that is close to the problem is more likely to respond effectively. Thus dispersion of the applied research function makes sense. The dispersion of the basic researchers along with their applied colleagues, as in the land grant universities, allows both types to take advantage of the knowledge externalities available due to close informal contact. The experience of institutions set up with a more exclusively applied focus, such as the International Rice Research Institute, apparently has led them to an increasing appreciation of a permanent, in-house, more basic research capacity.

Concentration on local problems also reflects the fact that their solution receives the greatest political support from local agricultural interests. Yield increases and cost reductions supplied gratis to all producers in an industry tend to reduce output prices rather than increase profits. But to the extent that the effect is only on local producers, the price reduction response is muted, and the local benefits to the sector are more likely to be positive. While productivity increases offer greater benefits if they occur on a national or international scale, the benefits would tend to go to consumers, who have little influence on the system. The result is that local problems get the most attention, and the spatially decentralized research system is well suited to addressing them.

The permanence of the research and extension staffing means that there is an accumulation of institutional capacity in the form of knowledge and expertise to respond quickly and effectively to emergency problems, such as the poinsettia whitefly or the abruptly-apparent selenium toxicity to waterfowl at Kesterson reservoir in California, as they arise. This "option value" could be important to the extent that the same response cannot be had as efficiently from the private sector in the form of temporary consultants or contractors. When the whitefly struck California, would it have been better for each affected farmer to have sent out for bids from private fly-problem-solvers?

The argument for public provision of quick access to a standing capacity for flexible emergency response seems similar to the argument for a publicly supported standing army or fire brigade (granted some would argue against the latter). The argument has force if in-house performance incentives are more appropriate, if the externalities from easy contact with and access to basic researchers are important, and/or if it would be difficult to know what contractor to choose if the expertise were not already present in the public sector.

The association of research with teaching is a practice that is widespread across the academic spectrum. As Ruttan (1982, p. 110) reports, "Over time, a consensus seems to have emerged in the United States that research is highly complementary to graduate education, but less so to undergraduate teaching." But the interplay between functions is difficult to analyze and not well understood. Obviously class time competes with research time, for faculty and students. On the other hand, the functions are complementary; in a sense each offers positive externalities to the other.

Students who learn how to apply their classroom learning by participating in real research in a critical environment under the supervision of their professors can reap educational and motivational rewards. Furthermore, their work has an actual social contribution, in contrast to fictional educational exercises. Their experience might also help students make better and earlier choices about the direction of their careers. Such benefits would normally become more available as the student advances in his or her academic career.

For professors and other teachers, involvement with institutional research helps keep their teaching relevant to current problems. This is likely to be more important for advanced undergraduate and graduate classes where there is usually more discretion about choice of subject matter and teaching tends to be more focused on research challenges. As researchers, their involvement in teaching, especially in advanced courses, helps broaden their perspective beyond their currently pressing research challenges to comprehend current work in other corners of their academic field, and in related specializations. Since scientific progress often results from drawing links between lines of investigation, involvement in teaching can encourage faster progress in research.

The above discussion has focused on some rationales for the current structure of agricultural research, as seen in the land grant system in particular. The features noted have their drawbacks, of course. Decentralization means inevitable duplication of some research (especially basic research) and of teaching functions. Permanent employment on a program basis makes it possible for "deadwood" to accrue and for the institutional culture to tolerate sloth and lack of responsiveness to social demands in both education and research. Teaching demands can divert bright minds from vital research tasks, and, on the other hand, research demands are currently being blamed for neglect of undergraduate teaching in the universities and colleges in general.

The social optimality of the land grant approach to agricultural research in trading off the advantages and disadvantages of its institutional design has not been scientifically established, of course. But its contribution to American agricultural productivity is well recognized, see, for example, Nelson and Wright, 1992. The relevant question now is how the existing structure of public and private collaborative research will respond to changes in the social, institutional and scientific environment .

Private and Public Sector Responses to the New Environment

We have some evidence already about the private sector response to the new opportunities. There has been an explosion in the private creation of new varieties after they were covered by the PVPA (Evenson 1983), and this occurred with conventional technology; it was not caused by the new possibilities associated with genetic engineering (U.S. Congress, Office of Technology Assessment, 1992). One might have anticipated this private sector response from the history of successful private production and marketing of hybrid corn varieties, which had some natural protection from unauthorized duplication by customers.

New advances in biotechnology have opened up a whole new technological frontier, and patented life forms and other genetic engineering products are already being marketed to agricultural producers as well as to other industries including prominently those in the health sector. Furthermore there is a complementarity between theinstitutional and technological advances. Modern analysis of DNA is likely to make policing of life form patents more effective.

The ability to patent and copyright has also changed the marketing possibilities for public and non-profit research institutions and the researchers who are employed in them. Whereas previously they had few opportunities to sell their output (as distinct from their services as research inputs) the institutions, and their employees, now face very significant rewards for success in meeting market needs, the diversity of which is reflected in the fact that two of the most successful to date are the Cohen-Boyer gene-splicing patent and Gatorade. An agricultural example is the domination of the market for strawberry varieties by the University of California, Davis.

Vertical Integration of Public Agricultural Research

The new opportunities to sell the property rights to embodied research outputs will affect the public and private research structures in many dimensions. Perhaps the most obvious is that private for-profit applied research is more feasible for these new innovations, so that the public role need not be vertically integrated right down to the farm gate, as it has been for other agricultural innovations without capturable property rights. Somewhere between basic research and extension, an interface can develop between the public and private innovation institutions. The transfers will tend to be vertical, with the private party downstream. If the transfer happens at the pretechnology stage, before the knowledge is embodied in a marketable product, it is similar to the public provision of technology to farmers, in that the private party acquires a free good. In this case, though, it is an input to further (private) research and development, rather than directly to the production process itself. This distinction can be crucial.

The purchaser(s) of university research output are likely to be corporate entities with substantial market power, not competitive farmers. To the extent that the clientele of university research is dominated by large powerful firms, administrators of land-grant institutions may have a "potentially massive public relations problem" (Kenney et al 1982 p. 52).

As noted above, market power in the relevant final product is essential where developmental expenditures are significant and any results are not protected by patents. If a potential purchasing firm is unprotected by pre-existing market power, it might well be reluctant to invest in the development of the technology to the marketing stage, for fear that others equally free to acquire the public technology gratis might beat it to the punch, or even copy the technology if it is too applied to pass the novelty and non-obviousness tests required for patenting.

The more novel the innovation, the less likely the availability of pre-existing market power to protect it. This might explain why the Commonwealth Scientific and Industrial Research Organisation in Australia found that they literally *could not give their technology away*. A policy of exclusive licensing was adopted to elicit greater interest in adoption of its discoveries by the private sector. Where this consideration is important, the public/private interface will tend to lie beyond the stage at which the first property right is acquired. Significant patenting will occur in the public part of the research sector. Private participation will replace some public efforts at the applied end of the research spectrum. This is already happening in other technological areas such as irrigation, where the dealers are the final agents of information transfer to farmers; a major part of extension ends at the dealer's yard. But substitution of private for public research will remain concentrated in the development stage, where further patenting is a possibility.

Another obstacle to direct technology transfer, found by Postlewait, Parker, and Zilberman in a survey, is the reluctance of in-house research departments to encourage the purchase of technologies that were not developed in the company itself. As a result, licenses to some of the most advanced technology developed in the U.S. have not been purchased by local companies. For example, a Stanford researcher invented an music chip for electronic keyboards. Despite the technology's obvious potential to revolutionize the industry, no American company wanted to license the chip, and eventually Yamaha licensed the technology and dominated the market.

Where the private innovator has market power, its research may, as mentioned above, extend up towards basic research, even without the legal protection of property rights. In this case the innovator may well be a large firm with a structure of bureaucracy possibly similar to that of a public institution.

Some other large firms take the opposite tack, acquiring technology by purchasing small companies that were developed around a certain innovation. Some large chemical companies lurk around trying to absorb promising innovative companies. In turn, these young companies need the marketing capacity of the big companies and they may seek an adopting parent. In effect some of the big companies are *marketing* organizations that rely on small R&D companies to develop a diversified product mix. They may also be potential customers for university research rights.

Increasingly, extension personnel are becoming more involved in giving policy advice to government and to public agencies, and in the facilitation of environmental management and controls. In these roles, they extend knowledge produced by university research. As the downstream reach of extension is rolled back in some areas of technology, it is expanded in other areas to meet changing needs.

University Marketing Arrangements

The possibility of patenting research findings in a public institution such as a land grant university raises many issues, among which are:

- How will the rights be marketed?
- Who shares in the revenue?
- Should the university participate in development investment?

Answers to some of these questions already exist (at least provisionally) in the structure of the "Office of Technology Transfer" (OTT) or of the "Office of Technology Licensing" (OTL), themselves institutional innovations seen in several universities.

The leader is Stanford, whose OTL is available for patenting and licensing the research of any faculty who wish to use it. The proceeds are divided as follows: After 15 percent is taken off the top to finance the OTL, net royalties are split into one-third shares for the inventor, one third for the inventor's department, and one-third for the university. In fiscal 1992, Stanford received $25.5 million in royalties and fees. (Barnum 1992)

The University of California has a similar systemwide office that awards university employees on a sliding scale, with 50 percent of the first $100,000 of net royalties, 35 percent of the next $400,000 and 20 percent of any higher amounts going to the inventor. Total revenues to the University from patents and royalties were $28.8 million in 1992 (Barnum 1992). In contrast to Stanford, faculty at the University of California, which has a central OTL, must use university services to patent university research. Some campuses, including Berkeley, are now developing their own campus-based OTL's to offer better service to their faculty.

Thus researchers at both public and private universities can stand to gain a substantial share of the realised value of their discoveries , and their departments and the whole institution also stand to gain. Paradoxically, *the explicit incentive appears greater in these public institutions than in the typical large private firm*, where the patents of employees are routinely assigned to the firm via prior contractual commitments and there is usually no significant explicit reward to the patent recipient.

What has been created is a monetary market for those types of innovation output that can obtain legal protection, within the context of the hierarchical bureaucratic structure of the university. Given the current popularity of markets as allocators of resources, the potential significance of this institutional innovation, for the university as well as for the researcher, should need little elaboration.

Some universities are now moving downstream again beyond patenting to financial participation in development of their patented technology, either directly or through a related institution to avoid legal problems of product liability. The University of California, for example, is considering the formation of California Technology Ventures Corporation to help commercialize the products of University research. It is time to question whether the university is an appropriate institution to handle the challenges and risks of participation in venture capital investment. Private inventors are notorious for having exaggerated views of the financial prospects of their brainchildren. In at least one case investment in venture capital has reportedly placed the financial health of a major private university at risk.

Beyond Patents to Partnerships?

The value of patent revenues in no way captures the contribution of university research to industry. Most of the important new biotechnology companies were created by university professors who linked up with venture capitalists to form new companies that developed and marketed new products. The founders of Genentech, Amgen, and Chiron, for example, include professors at Stanford, U.C. San Diego, U.C. Berkeley and U.C. San Francisco. Some of the top agricultural biotechnology companies have similar origins. The founders of Calgene and of Biosys include professors at U.C. Davis, and at U.C. Berkeley and Stanford, respectively. Generally the founders continue to be university professors while being involved with these companies.

One benefit of these companies is that, by their proximity and their personal links to the university, they are often good sources of employment for students and graduates. They also enable the university to continue to employ high-quality research professors while paying them less than the market value of their services.

But can the university design contracts that give it a greater share of the wealth which it helps to produce, off-campus, via its indirect contribution of prior research, expertise and other services? This is a challenge for the future, not unlike the challenge of optimal design of contracts for university research undertaken for the private sector. In both cases there are real pitfalls, including the danger of exposure of the university to legal actions (as seen previously in the tomato harvester case) and the danger of distortion of the university's research mission by private interest that "free ride" on university research efforts. (See Ulrich, Furtan and Schmitz (1986) for a discussion of free-riding by private-sector brewing companies on public research in Barley in Canada.)

Implications for University Performance

Research Efficiency

As noted, the frontier technologies we have been discussing happen to offer unusual opportunities for market returns due to patent and copyright protection. Patents and copyrights are very effective at encouraging the researcher to use his or her own information, informed by market pressures, to choose between research topics according to his or her capabilities, research costs, the probability of success, and the value if successful. Since research resource management is characterized by uncertainties and informational shortages, this utilization of the researcher's information and his or her market expectations is extremely important. The disclosure mandated under patent law also makes the information discovered more accessible for other members of society who can use it in further innovation efforts.

If, instead, a prize (money or promotion) is the reward for achieving a pre-specified goal, the researcher's information about the market value of success is unused unless an effective means of gathering it is found by the prize-setter. This does not matter if the latter has accurately identified an appropriate social goal. Some of the most important technical advances have occurred in response to prize incentives, including the technique of food preservation by canning, and the navigational chronometer. But in research an important part of the individual's skill is often the ability to know what questions to ask, what goals to set, given the economic environment and the technical possibilities, as set out in the (as yet incomplete) theory of induced innovation. (See for example Binswanger and Ruttan, 1978). Prizes for achievements defined ex ante do not reward such skills. If we assume the prize setter has similar skill and the latitude to use it, there may be no problem, but this is a big assumption.

On the other hand, if research contracts are awarded by competitive bids (an increasingly popular trend), private information about capabilities and success prospects is also lost to management.

When the research process is managed by central direction of research inputs including personnel, all of the private information about capabilities costs, probabilities, and market returns may be neglected. (For more on this see Wright 1983, 1985.)

But patents, copyrights and similar awards have their problems as research motivators. The race to be first to patent may involve wasteful duplication of effort on similar projects by personnel within or between institutions, especially if the resources devoted to a given line of research are very responsive to economic incentives (Barzel, 1968, Wright 1985). Duplication is made more likely by the need for secrecy about research strategies in preserving a competitive advantage. Collaboration with complementary research colleagues may be discouraged for the same reason. This problem will be particularly severe in large teams such as a research laboratory where individual contributions are difficult to verify.

In addition, the patent incentive might be too powerful in the sense that it distracts attention from other important tasks with less direct motivations. For university personnel, these could include teaching, advising and other institutional services, on the one hand, and research (such as more basic investigations) which yields non-patentable knowledge.

Similar kinds of objections to providing value-based rewards for innovation are increasingly expressed in the business management literature, largely influenced by recent Japanese thinking associated with the "Kaizen" (gradual improvement) system. They may also explain the observation that large private firms in the United States generally choose less high-powered, more implicit, rewards for their employees than the arrangements now becoming popular in research universities.

In the case of universities, it should be borne in mind that many of the problems with the new incentives, including duplication, envy, and misdirection of effort already exist in the system of rewards based on implicit criteria imposed ex post by deans and/or academic peers, from tenure and merit increases to general prizes such as the Nobel Prize, reflected in the adage "Publish or Perish." The advent of a parallel system of market-determined rewards might to some extent offset the distortions of the traditional implicit incentive structure. For example, the fact that researchers worry that the patent incentive biases research toward applications (Blumenthal et al. 1986 p. 1364) might be good news to those like Ed Schuh who claim that university research has lost a sense of relevance (Schuh 1986, 1991). In principal, this issue should be amenable to empirical resolution for specific cases.

Social Externalities

The market transfer of knowledge has been emphasized above. Two points are worth bearing in mind about the associated social contribution. First, an innovation, even if patented, usually transfers benefits to society greater than what the consumer pays. An innovator will often reduce, directly or indirectly, the price of some consumer goods, generating consumer surplus. Furthermore, the disclosure inherent in the patent process furnishes a knowledge externality, as mentioned above. In short, monetary returns do not necessarily constitute adequate rewards for invention, even in some cases where a strong patent is obtained. In these cases, public employment of researchers, and/or other incentives such as prestige might be beneficial.

Second, it would be a grave mistake to conclude that without the recent innovation in biological research property rights the university research contribution to private research activity would be negligible. As Nelson (1986) concluded from a 1984 survey of research managers by Levin et al., the role of university research is especially important in biologically-based industries (p. 187). More generally, "university research rarely in itself generates new technology; rather it enhances technological opportunities and the productivity of private research and development, in a way that induces firms to spend more both in the industry in question and upstream" (p.188). This stimulation is at least partially local. As Jaffee (1989) and Acs et al. (1992) show, states with high university research expenditures also have more industry research expenditures, more patents, and more reported innovations. The locations of Silicon Valley, the Route 128 area near Boston, and the emerging biotechnology industries near Berkeley and Davis support this view. As argued above, the spread of university patenting of life forms should expand this complementarity, especially in biological applications including agriculture, while also increasing the direct role of universities in the generation of new technology

Cash Cows?

As we have seen, the sale of research products can be a multimillion dollar enterprise for universities. The funds can help retain productive researchers who might otherwise go to industry, and can also augment the university' resources. But one should keep a realistic view of the possibilities here. Stanford is singularly successful in this research marketing area. Yet Stanford gets only about 11 percent of its budget from industry (Postlewait, Parker and Zilberman).

Conclusions

The agricultural research system is facing fundamental changes in the nature and complexity of its challenges and its opportunities. We can expect that this will result in less vertical integration of public research in several areas including production of new plant varieties, leaving a greater role for the private sector in applied research.

In other areas, the role of university research and extension may well expand. Many of the coming social demands can be met only if complex innovations are achieved in institutions and policies. The university, in addition to furnishing new technologies, should facilitate the debate on options and help shape the necessary institutional adaptations. In informing people about alternative risky choices, for example pesticide versus irradiation to preserve foods, the university should exploit its educational role in teaching and extension, as well as its research capabilities.

The new opportunities bring with them new management challenges. The potential for private gains from research property rights must be handled carefully. Its introduction of market signals for researchers will be a very positive development if it is not allowed to distract them unduly from teaching, advising and collaborative research activities with a less direct financial reward. To ensure that the latter does not happen, careful research is warranted regarding the structure of license-sharing arrangements and the determination of relative research contributions from collaborative projects.

Endnotes

1. Clear statements of the broad objectives of United States agricultural policies are surprisingly difficult to find. For our purposes, the following extract from the Hatch Act of 1955 is relevant:

   > "Sec. 2. It is further the policy of the Congress to promote the efficient production, marketing, distribution, and utilization of products of the farm as essential to the health and welfare of our peoples and to promote a sound and prosperous agriculture and rural life as indispensable to the maintenance of maximum employment and national prosperity and security. It is also the intent of Congress to assure agriculture a position in research equal to that of industry, which will aid in maintaining and equitable balance between agriculture and other segments of our economy." (U.S. Congress, 1980, p. 18-22)

2. For an excellent overview of prospective applications of genetic engineering of plants for control of insects, weeds, and diseases, of animals for growth promotion, animal health, reproduction technologies and creation of transgenic animals, and in food processing, see U.S. Congress, OTA (1992), and also parts of U.S. Department of Agriculture, Economic Research Service, (1992), especially Gibson (1992).

3. For an up-to-date discussion of relevant intellectual property protections, see Chapter 15 of U.S. Congress, Office of Technology Assessment, (1992).

References

Acs, Z. J. et al. 1992. "Real Effects of Academic Research: Comment." *American Economic Review*, 82.1, 363-367.

Alston, J. M., and P. G. Pardey. 1992. Technical Change, Constraints and Incentives: The Damaging Consequences of Getting Policies Wrong. Paper presented at the Conference "AgriculturalTechnology: Current Policy Issues for the International Community and the World Bank" organized by the Agriculture and Rural Development Department of the World Bank, Airlie House, Virginia on October 21-23, 1991. Davis, Calif.; St. Paul, Minn.: University of California, Department of Agricultural Economics; University of Minnesota, Department of Agricultural and Applied Economics, 66.

American Economic Association. 1985. "Papers and Proceedings of the Ninety-Eighth Annual Meeting of the American Economic Association." *Ninety-Eighth Annual Meeting of the American Economic Association*. The American Economic Review. Ed. Harvey S. Rosen and Wilma St. John. New York City. American Economic Association, 458.

Barnum, A. (November) 1992. "U.C. Wants More Research Profits." *San Francisco Chronicle*, 1992, C-1, C-3.

Barzel, Y. (August) 1968. "Optimal Timing of Innovations." *Review of Economics and Statistics*, 348-55.

Binswanger, H. P. and V. W. Ruttan. 1978. *Induced Innovation*, Baltimore, MD.: The Johns Hopkins University Press, 1978.

Blumenthal, D. et al. (June) 1986. "University-Industry Research Relationships in Biotechnology: Implications for the University." *Science* 232, 1361-1366.

Busch, L. (September) 1992. "Agricultural Reset in A Time of Change." Presented at Agricultural Research Institute's 1992 International Conference of Agricultural Research Administrators, McLean, VA.

Busch, L., W. Lacy, J. Burkhardt, and L. Lacy. 1991. *Plants Power and Profit*, Cambridge, Mass.: Basil Blackwell, Inc..

Cochrane, W. W. *Farm Prices: Myth and Reality*. 1958. Minneapolis, Minn.: University of Minnesota Press.

Evenson, R. E. (December) 1983. "Intellectual Property Right and Agribusiness Research and Development: Implications for the Public Agricultural Research System." *American Journal of Agricultural Economics*, 967-975.

Evenson, R. E., and B. Wright. 1980. *An Evaluation of Methods for Examining the Quality of Agricultural Research*. Paper. 6. Washington, D.C.: Office of Technology Assessment, U.S. Food and Agricultural Research, Work Group II. 73.

Gibson, D. (May) 1992 "Biotechnology Products." In *Research Progress in 1990-91*. U.S. Department of Agriculture, Agricultural Research Service, Washington, D.C., 28.

Huffman, W. E. and R. E. Just. (August) 1992. *Structure, Management, and Funding of Agricultural Research in the United States: Direction and Likely Impact. Report*. Ames, Iowa; College Park, Md.: Iowa State University; University of Maryland, 30.

Jaffe, A. B. 1989. "Real Effects of Academic Research," *American Economic Review*, 79.5, 957-970.

Kenney, M. et al. (July) 1982. *Genetic Engineering and Agriculture: Exploring the Impacts of Biotechnology on Industrial Structure*. Industry-University Relationships, and the Social Organization of U.S. Agriculture. Bulletin Series. 125. Ithaca, N. Y.: Cornell University, Department of Rural Sociology, 94.

Moffat, A. S. (May) 1992. "Plant Biotechnology: High Tech Plants Promise Bumper Crop of New Products." *Science*, 28, 770.

Nelson, R. R. and G. Wright. (December) 1964. "The Rise and Fall of American Technological Leadership: The Postwar Era in Historical Perspective." *Journal of Economic Literature*, XXX, 1931-1964.

Norton, G. W. et al. (August) 1992. *Economic Issues in Agricultural Research Priority Setting*. Report. Blacksburg, Va.; St. Paul, Minn.; Davis, Calif.: Virginia Polytechnic Institute and State University; University of Minnesota; University of California, 13.

Parker, D. and D. Zilberman. (February) 1991. *The Political Economy of Alternative Mechanisms of R & D Management in Agriculture*. Report presented to the Maryland Biotechnical Institute, Berkeley: University of California, 27.

_______. (Spring) 1992. *Biotechnology in a Regulated World*, Report. Berkeley: University of Maryland, Center for Public Issues in Biotechnology; University of California, Department of Agricultural and Resource Economics, 23.

_______. (February) 1992. *The Distributional Effects of Alternative Mechanisms of R & D Management*, Report. Berkeley: University of California, Department of Agricultural and Resource Economics, 29.

Postlewait, A. et al. (Fall 1991). *Technology Transfer and Biotechnology in the Bay Area*. Report, Berkeley, Calif.; College Park, MD: University of California, Department of Agricultural and Resource Economics; University of Maryland, Center for Public Issues in Biotechnology, 22.

Rose-Ackerman, S. and R. Evenson. (February) 1985. "The Political Economy of Agricultural Research and Extension: Grants, Votes, and Reapportionment." *American Journal of Agricultural Economics*, 67, 1-14.

Ruttan, V. W. 1982. *Agricultural Research Policy*. Minneapolis Minn.: University of Minnesota Press.

Science for Agriculture. (October) 1982. Report of A Workshop on Critical Issues in American Agricultural Research, June 14-15, 1982 Winrock International Conference Center, Petit Jean Mountain, Morrilton, Arkansas. New York, N.Y.: Rockefeller Foundation, 36.

Schuh, G. E. 1986. "Revitalizing Land Grant Universities, It's Time To Regain Relevance." *Choices*, Second Quarter, 6-10.

Schuh, G. E. 1991. "The Future of Land Grant Universities: Implications for Departments of Agricultural Economics." Presented at National Management Workshop for Agricultural Economics Department Chairs, Denver, Colorado, October 4-7, 1991. Minneapolis, Minn.: University of Minnesota, 24.

Sell, C. R. and M. A. Weiss. (May) 1992. "Pesticide Residues." In *Research Progress in 1990-91*. U.S. Department of Agriculture, Agricultural Research Service, Washington, DC, p. 24.

Takeoff, G. R. (May) 1992. "Food Fraud." In *Research Progress in 1990-91*. U.S. Department of Agriculture, Agricultural Research Service, Washington, D.C., p. 26.

U.S. Congress, House of Representatives, Committee on Agriculture. 1980. *Compilation of Statutes Related to Agriculture and Forestry Research and Extension Activities and Related Matters*, Hatch Act. 96th Cong., 2d sess. Washington, D.C.: Government Printing Office. Committee Print, 18-22.

U.S. Congress, Office of Technology Assessment. (August) 1992. *A New Technological Era for American Agriculture*, OTA-F-474. Washington D.C.: U.S. Government Printing Office.

U.S. Department of Agriculture, Agricultural Research Service. (May) 1992. *Research Progress in 1990-91*, Washington, D.C., 35.

Ulrich, A.; H. Furtan; and A. Schmitz. (February) 1986. "Public and Private Returns from Joint Venture Research: An Example from Agriculture." *Quarterly Journal of Economics*, 101, 103-129

Wade, N. "Gold Pipettes Make for Tight Lips." (June) 1981. *Science*, 212, 1368.

Warrick P. (August) 1992. "The Real Mothers of Invention." *San Francisco Chronicle*, pp. D3-D4.

Wright, B. D. 1983. "The Economics of Invention Incentives: Patents, Prizes, and Research Contracts." *American Economic Review*, 73.4, 691-707.

Wright, B. D. 1985. "The Resource Allocation Problem in R & D." *The Economics of R & D Policy*. Ed. George Tolley. New York, N.Y.: Frederick A. Praeger, Inc., 41-67.

Chapter 21

Strategies for Change: Agricultural Research in the U.S. Public Universities

William B. Lacy
The Pennsylvania State University

Agricultural research in U.S. public universities faces new challenges as a consequence of changes occurring both within and outside the community. The broader organizational environment confronting public universities and, in particular, colleges of agricultural sciences is changing rapidly. These changes have been documented by a number of individuals, including several authors in this volume. Among the leading political and social factors affecting the context in which public universities conduct research and graduate education are the increasing globalization of all sectors of society, a large federal deficit and current economic slowdown affecting both state and federal budgets, changing state and national populations, the potential for global climate change and increasing environmental concerns, and declining public confidence in government and science.

In addition to these general contextual trends, specific activities at the federal and state level are or will impact the research activities of U. S. universities. The federal government has been the major source of support for university research in the United States during the last several decades. The National Science Foundation and the National Institutes of Health currently fund approximately $6.5 billion of university research and several other government agencies, including the Office of Naval Research and the Department of Energy, spend several billion additional dollars (Marshall and Palca, 1992). However, federal science policy is currently undergoing a comprehensive assessment. The House Science and Technology Committee is reviewing federal research programs to determine whether they address a broad range of social goals. Representative George Brown, chair of the committee, has indicated his unwillingness to exchange the cold war military rationale for strong support of science for a single new rationale of promoting industrial competitiveness. Instead, he supports development of a set of national goals as a basis for the federal science policy (Cordes, 1992a).

At the state level, legislators and governors are calling for increased efficiency and continued budget reductions for public funding of research in the 1990s. Higher education no longer enjoys a privileged place in the state budget process. Although thirty-five states cut their previously adopted fiscal year 1992 budgets, many protected certain spending categories, such as welfare programs, corrections, and primary and secondary education. Higher education was rarely exempted from these budget cuts. Consequently, many predict that higher education will never return to the prosperity of the mid-1980s. Moreover, legislators may continue to pressure faculty to be more productive (often translated as heavier teaching loads) and more accountable to the public (Yudof, 1992; Lively and Mercer, 1993).

At the same time, within research universities major changes are being contemplated, potentially as fundamental as the great transformation that swept higher education more than a century ago. That transformation merged the German model of a university primarily devoted to the advancement of knowledge through scholarship and research and the American Populist model of the land-grant system which held that higher education should serve not an elite class but qualified young people from all walks of life. Today many university presidents and other prominent academic leaders argue for the development of a new institutional paradigm or model as we move into the twenty-first century (Grassmuck, 1990).

Among the challenges, opportunities and conflicting goals confronting land-grant research universities are: (1) improving undergraduate education and achieving a better balance between research and teaching; (2) globalizing campus programs and internationalizing the student body; (3) balancing efforts to promote openness in scientific research and international collaboration with the demands by industry and government to tighten intellectual property laws and pursue national preeminence; and (4) restructuring all areas, academic and administrative, to deal with the tough financial problems facing higher education and particularly public higher education (Zinberg, 1991). An American Council on Education report indicates that the poor economy has taken a heavier toll on public universities than on private ones. In a survey of more than 400 universities, 47% of the public 4-year colleges and universities, compared to only 14% of the private universities, had unchanged or declining budgets in 1992 (Marshall and Palca, 1992).

Within this current context, aggressive strategic planning, creative and innovative action, and a strong commitment to excellence on the part of colleges and universities will be essential. Additionally, land-grant universities will need to explore and utilize a variety of organizational structures, including interdisciplinary centers, institutes, and laboratories, as well as intercollege and interuniversity activities, in order to pursue their research agendas. Several examples of innovative institutional actions and strategies employed to promote agricultural research are examined in the remainder of this paper. They are organized into four broad categories: college; intercollege; interuniversity; and interagency or interorganizational (university-government-industry-private nonprofit). While these examples do not exhaust the rich diversity of strategies being developed, they do reflect a range of these efforts.

College Strategies

In U.S. public universities, colleges of agriculture and natural resources have been engaged in agricultural research for over 100 years. During that time they have changed dramatically, becoming increasingly specialized around disciplinary departments and a reductionist approach to science (Busch and Lacy, 1983). While these changes have resulted in numerous major breakthroughs, creative discoveries, and new inventions which have increased agricultural production and productivity, they have also contributed to a narrow research agenda defined by the dominant agricultural disciplines and to the de-emphasis of complementary systems-based disciplines that are needed to address the complex issues facing our food and fiber system. Often neglected areas of disciplinary research have included ecology, integrated plant, animal and food systems, and social science and nutrition research. Hiring new scientists from a wider range of disciplines, already being implemented selectively, will need to be expanded and evaluated for future hiring policies. In addition, to bring new perspectives to the system, attracting scientists from diverse institutions may be equally important. In the past, nearly all agricultural scientists have come from a small number of large, land-grant colleges of agriculture. Although this has resulted in a community of scholars with a shared sense of mission and goals and a close network of colleges throughout the system, it has often led to insularity and isolation of agricultural scientists from the broader scientific community. The strength of this community and the strong network among scientists should be maintained, while attempting to seek scientists from diverse disciplines and institutional backgrounds and experiences.

Colleges also need to consider additional institutional mechanisms, particularly organizational structures that complement the disciplinary structure for addressing the new and broader agendas. Many institutions have explored establishing interdisciplinary committees, graduate research programs, and centers within their colleges. When organized properly, these mechanisms can stimulate interdisciplinary collaboration, provide focal points for dialogue and discussion of these issues, and serve as stimuli for seeking extramural funding. Recently colleges of agriculture across the country have established centers within their colleges to focus on topics such as bioremediation, biodiversity, biotechnology, economic and community development and wetlands. Leadership, funding and institutional support are key criteria for the success of these centers. Without them these centers may become little more than letterhead organizations.

Establishment of interdisciplinary graduate research programs that enable the next generation of scientists to emerge with a strong disciplinary foundation, as well as a broader understanding and appreciation of the knowledge and issues being raised by a number of other scientific perspectives, is another promising effort. While such programs have existed in the past, new programs should stretch the disciplinary boundaries and provide a better integration of diverse perspectives. These programs should complement, not replace, the strong disciplinary efforts. Some of the more recent innovations in this area have included graduate programs in ecology, plant physiology, and sustainable agriculture. These programs, however, often lack adequate institutional support and must rely on academic departments for resources and faculty time.

Interdisciplinary research and education task forces or issues committees are a third intra-college effort being explored by a number of colleges of agriculture to strengthen both their research and educational capacity. The responsibilities of these committees often include identifying and analyzing needs within the issues areas; defining appropriate action strategies and recommending mechanisms for implementing these strategies; communicating the needs, strategies and other information regarding the issues to appropriate college and university faculty and staff; and interfacing with foundations and appropriate federal, state and local agencies. The twelve issues committees established at Penn State's College of Agricultural Sciences to address the key issues of integrated animal management, integrated crop management, improved human nutrition and health, food safety, water quality, management of natural resources, economic and social viability of communities, family social and economic well-being, development of human capital, animal health, sustainable agricultural systems, and solid waste management are one example. These committees are composed of members of the extension community from throughout the state, as well as researchers in the academic departments, and have been particularly effective in linking extension and research. As problems facing people at the community level become more complex and diffuse, these committees may form a mechanism for maintaining a strong research base to support extension education programs. These committees also foster and promote interdisciplinary networking within the college. They focus on the analysis of issues, not on the implementation of specific programs which remains the primary purview of the academic and regional units. The success of these committees is dependent on a number of key factors: including, defining their purposes clearly, ensuring the membership is appropriate, selecting leadership, providing modest funding, and ensuring that their activities do not become isolated from the college's other programs and activities (Hood, Schutjer and Evans, 1990).

Intercollege Strategies

As noted above, a major criticism leveled at public sector agricultural research has been its isolation. Approximately a dozen land-grant colleges of agriculture provide most of the graduate education for the system's new scientists and within individual universities communication across college lines often remains limited. To address the range of issues that these colleges are facing today and in the future, it will become increasingly necessary to draw upon the broader scientific community. Innovative mechanisms for linking scientists in colleges of agriculture with those in other science communities will be needed.

One effort to stimulate intercollege research programs has been to provide seed funds for collaborative research that involves scientists from the college of agriculture and from at least one other college. One such program provides $300,000 - $400,000 annually for an intercollege research program in three areas, the biological sciences, physical and engineering sciences, and the social and economic sciences. This program requires that scientists in the college of agricultural sciences collaborate with scientists in other colleges in the university. Proposals are peer reviewed and three to four grants are awarded annually. A large number of new collaborations has been stimulated among both funded and unfunded intercollege teams. The program also has led to numerous extramurally funded grants being awarded to these new intercollege research teams. In addition, it has enabled the college to focus university-wide expertise on its scientific issues and problems.

Structural changes, for example, joint or adjunct faculty appointments, can also ensure greater intercollege collaboration and cooperation. More permanent efforts include formation of intercollege departments, such as agricultural engineering departments within both colleges of engineering and colleges of agriculture, and rural sociology departments linked with departments of sociology in colleges of arts and sciences.

More recently, new intercollege research programs, institutes, centers and public corporations are being formed in universities to address a range of interdisciplinary issues and opportunities. A few of these organizational structures have existed for several decades, and, when compared to other university organizations, tend to be more diverse, better financed, and more aggressive in seeking extramural funds. They often involve large capital investments in buildings and equipment. In the area of agricultural biotechnology at least 17 state university research center initiatives are located at land-grant universities, including the New York Center for Biotechnology in Agriculture at Cornell University (1983), the Center for Agricultural Molecular Biology at Cook College, Rutgers University (1987), Iowa's Office of Biotechnology at Iowa State University (1984), and the

Biotechnology Institute at The Pennsylvania State University (1984). At Michigan State University a nonprofit corporation was established to develop and market innovations arising from research. While these centers and institutes offer many advantages in conducting research focused on particular problems, several concerns have been raised, including the nature of their research agendas and the divided commitments and loyalties created by joint appointments in a center and in an academic department (Lacy and Busch, 1991). Expectations for the success of these ventures run high. However, most of these programs are, as yet, too young to evaluate.

Merger of two or more colleges, an increasing likelihood at several universities, may be the most comprehensive intercollege effort. Already the situation in some smaller states, this approach has the advantage of creating a critical mass of scientists in related areas, such as agriculture, natural resources and forestry. However, following merger the research agenda may become more diffuse and the specific issues of agriculture may become less prominent. The colleges of agricultural sciences will need to take an aggressive leadership role in identifying appropriate linkages and mergers and in developing mechanisms to make these transitions successful.

Interuniversity Strategies

For many years the agricultural research community has been a leader in interstate university efforts. With the passage of the federal Research and Marketing Act in 1946, one-quarter of the formula funds (Hatch and McIntire-Stennis) were set aside for regional research, thereby stimulating a large number of interuniversity collaborative efforts. Today many regional research projects involve not only multiple states but multiple regions. In general, these regional research projects have been quite successful. However, they require more coordination and cooperation among scientists and more administrative support than individual scientist projects. In addition, they require travel to annual meetings to share research findings and to ensure meaningful coordination and collaboration. The additional effort and costs involved may dissuade some good scientists from participating in promising regional projects.

The regional rural development centers have stimulated another effective interuniversity research activity. These centers are located in four regions of the country and promote research and education projects and conferences of regional importance. However, these centers have been severely hampered by low level annual funding from the federal government. Indeed, federal appropriations for research programs through these centers have been half that of the appropriations for extension activities.

More recently, states with similar research agendas in particular commodities have begun exploring the possibilities of merging research efforts. For example, Cornell University and Penn State University have recently signed an agreement to eliminate duplication in their grape research programs and to jointly operate a grape research facility. In times of declining resources, this has enabled both states to maintain a quality research program. When these efforts are attempted, however, it is extremely important to maintain an open dialogue with the many clientele groups and interested parties traditionally served by separate state facilities. If the dialogue is successful, duplicative state research programs in neighboring states may be merged effectively.

Interorganizational Strategies

Increasingly, public research universities are expected to address social issues that have proven intractable to the larger society. For example, many current national, regional, state and local economic development policies presume that, with appropriate partnerships, the universities will provide scientific and technological breakthroughs, create jobs and fuel economic development. These financial and political pressures on public universities encourage closer university-government-industry ties and emphasize the commercialization of research for the development of new products and processes. While a rich tradition of university, government and industry linkages exists, the agenda has been modest and the proportion of the university budget devoted to these activities has been rather small.

Recent scientific developments in such areas as molecular biology, biotechnology, materials science and computer science, as well as federal and state financial incentives and political actions, are intensifying pressures for universities to work more closely with industry. For example, the National Science Foundation and the National Institutes of Health are adopting strategic plans that emphasize commercialization of the research they support. The executive branch of the federal government, concerned that U.S. based companies are losing their technological edge and their share of world markets, is encouraging universities to team up with businesses to speed the development of new products. Similarly, states, such as Pennsylvania (Ben Franklin Partnership) and Ohio (Thomas Edison Program), have funded economic development programs through research grants to universities based on university-industry collaboration (Blumenstyk, 1992).

Colleges of agricultural sciences have traditionally maintained working relationships with agricultural chemical, farm implement, and food processing firms. With the emergence of biotechnology, new types of college and/or university-industry relationships are developing which appear to be more varied, wider in scope, more aggressive and experimental in nature and higher in public visibility than the relationships of the past. In agricultural biotechnology, a number of university and industry relationships have been established. For example, Cornell University has specific grants from Agrigenetics Corporation (tomato hybridization through cell culture), Sandoz, (tissue culture generation), and Nestle's (bitterness in squash); Purdue University has grants from UpJohn and Dow Chemical (soybean research), Eli Lilly (wheat genetics), and Quaker Oats (oats research); and The Pennsylvania State University has arrangements with Hershey Foods, Mars, Inc., and Ambrosia Chocolate Company (molecular biology of the cacao plant). Other university-corporate relationships in agricultural biotechnology include the University of Arkansas with Busch Agricultural Resources, Washington University with Monsanto and the University of Illinois with Standard Oil. Perhaps the most complex university-government-industry consortium to focus on agricultural applications of biotechnology is the mid-west plant biotechnology consortium which involves 15 mid-western universities, 3 federal laboratories, 37 agribusiness corporations with headquarters in the mid-west, and research institutes from 8 states. The consortium's purpose is to conduct basic research and plant biotechnology and to promote the transfer of technology to foster the economic competitiveness of U.S. agriculture and agribusiness (Busch, et. al., 1991).

Another important example of university, government and industry collaboration is the new USDA program, Alternative Agriculture Research and Commercialization (AARC). This program seeks to facilitate and accelerate the development and commercialization of new industrial (non-food, non-feed) products manufactured from farm and forestry materials. An independent entity within USDA, the AARC Center conducts product and co-product research, and development and commercialization activities to enhance the marketability and economic viability of industrial products from agricultural materials. Proposals are reviewed for both technical and business merit (see paper in this volume by Paul F. O'Connell, AARC Center director).

The development of these relationships between university, industry and government may have both positive and negative consequences for the universities. University and industry collaborations may bring useful products to market more rapidly and promote U.S. technological leadership in a changing world economy. In light of funding stagnation at both the federal and state level, these collaborations may provide a means of raising new funds for university research. Finally, these joint efforts may expand the scientific network, thereby increasing communication between some industry and university scientists.

A number of concerns, however, have been raised regarding these new relationships. First, long-term research, previously a major emphasis in the public sector, may decline as a result of the private sector's emphasis on short-term proprietary goals and inclination to support projects which span one year or less. Any restriction of scientific communication or access to information is also cause for concern. Proprietary agendas have already begun to inhibit the flow of information among biotechnology scientists in the agricultural sector. This is particularly problematic for university scientists with private sector grants who often must delay public discussion or publication of their work or its results pending review by the sponsoring company (Matkin, 1990).

The increasing regulation of universities that accompanies the federal funding system is another important concern. Derek Bok, President emeritus of Harvard University, has urged policymakers not to overwhelm universities with regulations and requirements for relevance in the basic science they pursue, since fundamental breakthroughs cannot be scheduled (Cordes, 1992b). The government may attach numerous conditions to research funding, the manner of the distribution of the findings, and even the selection of the research itself. Moreover, some government officials are pressuring colleges to reconsider their ties to foreign students, companies, and governments. Recently, Robert White, Under-Secretary of Commerce for technology, stated that universities should be careful in establishing research relationships with foreign companies and that such ties should be allowed only if American researchers gain access to foreign scientific information. However, as G. L. Wood, Dean of the Graduate School at the University of Washington indicated, "Trying to keep secrets in the world of graduate education and research is a sure ticket to mediocrity. The idea that we can somehow build an information wall around this country is very short-sighted" (Jaschek, 1992: A34).

Perhaps the most serious concern raised about these linkages has been the potential for conflict of interest or scientific misconduct. At issue is whether such linkages between universities and industry will entangle faculty members in financial conflicts of interest and divert universities from teaching and basic research. Possible types of scientific misconduct include potential favoritism, unwarranted financial advantage through privileged use of information or technology partly derived from publicly funded research, constraints on sharing of materials and information, and shelving of research which may be of interest to the public but not to the corporation. In addition, some critics suggest that the government's failure to integrate scholarship related to other national goals, for example, social science research, with science and technology aimed at industry may result in public rejection of the work or cause unanticipated problems (Cordes, 1992b).

Recently, Derek Bok, in his final president's report to the Harvard University board of overseers, warned that commercialization of the university may be the most severe threat facing higher education. Mr. Bok noted that as universities become "more entrepreneurial, they appear less and less as charitable institutions seeking truth and serving students and more and more as huge commercial operations that differ from corporations only because there are no shareholders and no dividends." He concluded by saying, "It will take very strong leadership to keep the profit motive from gradually eroding the values on which the welfare and reputation of universities ultimately depend" (McMillan, 1991: A31).

Some analysts have suggested that in federally supported efforts to develop new technologies university officials should seek a much broader range of participants/partners. Besides industry and government representatives, these should include representatives of labor, environmental, community, health, and social service groups, as well as private foundations and philanthropic organizations.

An example of such a broad-based effort, is the Penn State/Rodale Center for Sustaining Agriculture and Natural Resources in Urbanizing Environments, a collaboration between a college of agricultural sciences and a private nonprofit foundation, in conjunction with support from federal agencies and the state. This new center, established in 1992 with support from the USDA's Cooperative State Research Service, Extension Service, and the Agricultural Research Service, is designed to help agricultural producers, policy-makers and communities develop environmentally and economically sustainable practices and polices at the rural/urban interface through programs in research, education, and public policy.

The USDA's Sustainable Agriculture Research and Education (SARE) Program is another unique example of new partnerships in the agricultural research endeavor. Initiated in 1988, the program is managed through four regional administrative councils composed of farmers and ranchers, and representatives of nonprofit private, agribusiness, federal and state government and academic organizations. Most funded projects include active involvement of farmers or ranchers. The program places emphasis on whole-farm systems research, on-farm participation, environmental and economic impact assessment and quality of life programs. However, the program has attempted to address far more than its limited budget ($6.7 million annually) realistically permits. Substantial efforts are underway by several alternative agriculture, environmental and consumer groups to enhance future federal funding for this program.

The most elaborate examples of university-industry collaboration are research parks developed on or near university campuses. Conceived about a half a century ago to link universities with the private sector, research parks have mushroomed in recent years with over two-thirds of the 115 U.S. and 307 worldwide parks being established after 1981. Since their inception, universities and private developers have invested $6.5 billion to erect 65 million square feet of research park space where nearly 190,000 Americans, including a growing number of research scientists, work (Goldstein and Luger, 1992). Designed with expectations for spurring economic growth throughout a particular state or region, most parks in the U.S. have a university connection and are directly or indirectly supported by state and/or local governments.

Based on the widely perceived success of early research parks, most notably the Stanford Research Park in California and the Research Triangle Park in North Carolina, new developments seek to attract high technology research and development companies and government laboratories that would benefit from proximity to a research university. Research parks tend to require a lengthy development period before they can begin to contribute to the restructuring of a region's economy. Successful parks have required excellent leadership, vision and patience on the part of the region's government, business, and academic leaders, as well as continuing financial support for a long time period. Some analysts note that even a strong research base and a commitment to industry are not enough. A university based research park must also have a good business address. For example, two Florida-based research parks have had different success, in part because of location. Although located near a major research university, the University of Florida, a park near Gainesville has not been developed nearly as rapidly as a similar one in Orlando, a popular business location (Kefalides, 1991). To date, most research parks have not been successful. In fact, while research parks are highly visible and can have significant symbolic value for a region, other types of technology development programs and university-industry collaborative efforts may be more cost effective.

Conclusion

Colleges of agriculture are facing wide ranging challenges to their traditional research agenda from an array of potential new clients and partners, as diverse as environmentalists, alternative agriculturists, and consumer advocates. As this paper has documented, the current institutional environment is producing partnerships between universities and state and federal governments, between the public and the private sector, between profit and nonprofit organizations and between national and international organizations, as well as interdisciplinary teams among a wide array of researchers and practitioners. It is also requiring new strategies and approaches to build these partnerships and interdisciplinary team efforts. Creating an expanded pluralistic community of stakeholders will not be easy. For U.S. public institutions to be successful in instituting innovations and new strategies for agricultural research, they must assess the goals and values of these different groups and their relationships to the mission and goals of the college of agriculture in which that research will be conducted. In addition, mechanisms must be established for developing consensus among the participants, since many of the new stakeholders will have quite different long-range goals, directions, and priorities.

Each of the strategies for coping with the changing social, political and financial context for research will require careful examination of both what the university is doing and what it should be doing. This will entail fundamental tensions, and, in the midst of these increasing and cross-cutting pressures and demands, maintaining a clear and focused perspective in charting future agendas will be difficult. Any strategies or innovations must be evaluated relative to the fundamental purpose of research and graduate education programs in public land-grant universities, creating new knowledge and opening minds through education and the exchange of that knowledge.

References

Blumenstyk, G. 1992. "States re-evaluate industrial collaborations built around research grants to universities." *The Chronicle of Higher Education*, XXXVII, 25, A1, A24, A25.

Busch, L. and W. B. Lacy. 1983. *Science, Agriculture and the Politics of Research*. Boulder, CO: Westview Press, 1983.

Busch, L.; W. B. Lacy; J. Burkhardt; and L. R. Lacy. 1991. *Plants, Power and Profit: Social, Economic, and Ethical Consequences of the New Biotechnologies*. Cambridge, MA: Blackwell.

Cordes, C. 1992a. "Broad review of U.S. science policy slated in House: Social goals cited." *The Chronicle for Higher Education*, XXXIX, 5, A29.

Cordes, C. 1992b. "Debate flares over growing pressures on academe for ties with industry." *The Chronicle for Higher Education*, XXXIX, 4, A26-A29.

Goldstein, H. A. and M. I. Luger. 1992. "University-based research parks as a rural development strategy." *Policy Studies Journal*, 20, 2, 249-263.

Grassmuck, K. 1990. "Some research universities contemplate sweeping change, ranging from management and tenure to teaching methods." *The Chronicle for Higher Education*, XXXVII, 2, A1, A29-A31.

Hood, L.; W. Schutjer; and D. Evans. 1990. "Integration of research and extension: It's a priority at Penn State." *Choices*, 5, 3, 22-24.

Jaschek, S. 1992. "Colleges are pressed to reconsider ties to foreign students and companies." *The Chronicle for Higher Education*, XXXVIII, 35, A27, A34.

Kefalides, P. 1991. "Proliferation of research parks conceals uneven success pattern." *The Scientist*, 5, 18, 1,6,7,17.

Lacy, W. B. and L. Busch. 1991. "The fourth criterion: Social and economic impacts of agricultural biotechnology." In J. F. MacDonald (ed.), *Agricultural Biotechnology at the Crossroads: Biological, Social and Institutional Concerns*. Ithaca, NY: National Agricultural Biotechnology Council, 153-168.

Lively, K. and J. Mercer. 1993. "Higher education and the states." *The Chronicle for Higher Education*, XXXIX, 18, A29.

Marshall, S. and J. Palca. 1992. "Cracks in the ivory tower." *Science* 257, 1196-1201.

Matkin, G. W. 1990. *Technology Transfer and the University*. New York: McMillan Publishing Co.

McMillan, L. 1991. "Quest for profit may damage basic values of universities, Harvard's Bok warns." *The Chronicle of Higher Education*, XXXVII, 32, A1, A31.

Yudof, M. G. 1992. "The burgeoning privatization of state universities." *The Chronicle for Higher Education*, XXXVII, 36, A48.

Zinberg, D. S. (ed.). 1991. *The Changing University: How Increased Demand for Scientists and Technology is Transforming Academic Institutions Internationally*. Boston, MA: Kluver Academic Publishers.

Chapter 22

Institutional Innovations and Strategies for Public Supported Agricultural Research

Paul F. O'Connell
AARC Center Director, USDA

The Agricultural Research Community needs to revisit it's roots if it hopes to obtain the support required to carry out their scientific studies. That support depends on user relevance, feedback mechanisms, awareness of emerging issues, and most important a true partnership with the final consumer of the research product. Many farmers, commodity groups, agro-industry and public interest groups cannot identify with the current research users in the selection, management, and implementation of research programs.

Clientele directed agricultural research has been put forth as a desirable model for other sectors of the economy to help in recapturing the nation's competitive edge (Nelson and Langlois, 1983). People with vision like John Gregory (University of Illinois) and Eugene Hilgard (University of California) established the State Agricultural Experiment Stations. Others led the way for the formation of the Cooperative Extension Service and the Agricultural Research Service.

Scientists and engineers came up with plants and animals that grew faster, resisted disease, and were more nutritious; soil management practices that enhanced production; labor-saving equipment; and farm management techniques that allowed producers to become better business persons. County Extension Agents transferred this knowledge to farmers with easy-to-read technical publications, demonstration projects, and one-on-one interactions.

This model worked well because there was a continual flow of information between scientists, extension agents, and farmers. When a new seed variety, piece of equipment, or soil conservation practice became available, it was tested on a farm with a progressive operator. If there were some flaws in the idea, the problems went back to the researcher for further study and correction.

The agriculture industry has matured, and in many parts of the country this integration among scientists, extension agents and farmers no longer occurs. To a large extent, commercial farmers now depend on the farm supply store or have their own consultants to provide guidance on what pesticides, seeds, and fertilizers they should purchase and how they should be applied. Over the last several years I have attended several meetings of the National Corn and Wheat Growers and found that they had little knowledge of State Agricultural Experiment Stations. This observation is not intended to suggest there isn't a relation between public supported research and today's farming practices. However, what I am suggesting is that many commercial operators do not make the connection. On the other end of the spectrum, an increasing proportion of agricultural research is basic or discipline oriented. Scientists are rewarded for grants they obtain, publication in prestigious journals and evaluation by their peers. This situation does not allow much room for the concerns of the final user.

Earlier in this volume, Donald Holt from the University of Illinois, discussed the linear versus parallel technical development models -- and the consequences of using each. Over the last several years the public supported agricultural research establishment has moved toward the linear approach and the result has been increasing separation between research results and practical knowledge. In my opinion, the parallel model is preferable because it encourages continued interactions among basic, developmental, and applied research.

In the last 5 to 7 years I have been involved in two major program areas that have tended more toward the parallel model. These are Sustainable Agriculture and Industrial Uses.

In FY 1988, the initial funds were appropriated for the sustainable research and education program (formerly know as LISA). The first step was to set up administrative councils and technical committees in each of the four traditional SAES regions. To encourage interactions among researchers and users, members of the councils represent producers, non-profits, scientists, and technology transfer specialists. An Experiment Station official coordinates the program in each region, but the decision on program priorities and competitive project funding is a joint decision of Administrative Council members. All parties take an active part in the process and thus have ownership in the program. The result has been a ground swell of mutual respect among scientists, non-profits and producers involved in the program (1992 GAO report on Sustainable Agriculture).

Another program where scientists and users are actively involved is in the Alternative Agricultural Research and Commercialization program, commonly known as AARC. It's purpose is to facilitate the commercialization of new industrial products that use farm and forestry materials. Policy is established by a nine person board with seven representing the private sector. A competitive effort is now underway where private companies are encouraged to submit proposals for funding where they share in the R&D costs. Proposal review is done by both business and scientific specialists. Opportunities exit in a wide array of markets from degradable polymers (plastic substitutes) and oxygenated fuels to lubricants, coatings and annual fiber crops that can compete with imported newsprint.

Both of these examples illustrate the advantage of close user involvement. There is a constant interaction between the final use of R&D and the scientists. There is a focus to their efforts as envisioned by the original leaders in agricultural research. With limited funds the shotgun approach to technology exploration is replaced by a more targeted effort. As stated by Bernadine Healy, Director of the National Institute of Health, in a 1992 article in *Science*: "NIH does not exist to do science for science's sake, but rather for practical and humanitarian purposes: to improve and preserve the health of the American people." A similar argument can be made in agricultural research.

References

Healy, Bernadine. 1992. "Is this your Father's NIH? and other Strategic Quotations." *Science*, Vol. 257.

Nelson, Richard R. and Richard N. Langlois. 1983. "Industrial Innovation Policy: Lessons from American History." *Science*, Vol. 219.

U.S. General Accounting Office. 1992. "Sustainable Agriculture: Program Management Accomplishments, and Opportunities." GAO/RCED-92-233.

Conclusions

Chapter 23

Strategic Issues Facing the U.S. Agricultural Research System

Robert D. Weaver
The Pennsylvania State University

An underlying tone common to the first section of this volume is that while past contributions of the U.S. agricultural research system to expanding productivity have been substantial, the system is currently under substantial stress and in a period during which substantial transition and change in the focus, scale, and scope of research can be expected. As viewed by these authors, a symptom of this stress has been waning societal support for the traditional productivity goals of both private and public sector agricultural research. Matching this trend, the authors cite increasing societal pressure for research which addresses what might be labelled the *qualitative productivity* implications of agricultural productions technologies. These implications include both the *externalities* of agricultural production as well as the qualitative aspects of agricultural output. Economists define externalities as outcomes of a producer's or consumer's activities that affect other producers and consumers. For example, if farmer A spreads nitrogen and it leaches into the groundwater, the impact on other producers and consumers would be an externality. Other externalities of interest have included soil erosion, water quality, and air pollution. Societal interest in the qualitative aspects of agricultural production has emerged over the past decade focusing on: e.g. nutritional quality, residues associated with food products, the technologies used in production and marketing, and the extent of processing. In each case, these aspects of agricultural production are difficult, if not, infeasible for the farmer to monitor, control, market, and personally benefit from.

Viewing this shift in the demand for agricultural research within the context of the historical motivation of the State Agricultural Experiment Station (SAES) system (see e.g. Marcus;1986), the shift is one which takes the focus of research away from purely sectoral goals such as quantitative productivity to more broadly defined societal goals of ensuring the achievement of maximum social welfare, or the greatest good for the most people. From another perspective, the shift suggests that society no longer accepts the logic that increased quantitative farm productivity necessarily contributes to increased social welfare. Finally, the shift reflects a trend that has affected research more generally, the disenfranchisement of the scientist as the director of the path of research and the reclaiming by society of that responsibility.

The extent to which societal demand for agricultural research has evolved to include the qualitative productivity goals is broadly acknowledged, as evidenced in the Joint Council on Food and Agricultural Sciences' priorities for agricultural research, teaching, and extension (Joint Council, 1992):

* Attain Sustainable Agricultural Systems Compatible with Environmental and Social Values.

* Develop Scientific and Professional Expertise to Advance New Technologies and Global Relationships.

* Address Critical Social Issues in a Changing Global Society.

* Enhance the Global Competitiveness of U.S. Agriculture.

* Provide a Safe, Affordable, and Nutritious Food Supply.

In the private sector, a further manifestation of the role of societal demands has been increased regulation of research (Weaver, in this volume) and increased demands for control of the agenda of public sector research through centralization of agenda formation and use of directed funding (BOA/NRC, 1989). Paralleling these trends has been a decline in public confidence in the traditional agricultural research system and a willingness and interest in broadening the scope of participation in the solution of the problems which are currently ranked as priority. In part, this issue has been debated since the founding of the Land Grant system (Marcus, 1986; Schweikhardt and Bonnen, 1986). However, the debate formally re-emerged to a national level in the early 1970s (The Pound Report NRC, 1972; Winrock, 1982; and NRC, 1985) as motivated by popular criticism (e.g. Carson, 1962; Hightower, 1978; Kenny and Kloppenburg, 1983).

The apparent stress on the U.S. agricultural research system leads naturally to the question of whether this condition is unique to agriculture, or instead whether it is a condition that affects research more generally. From a different perspective, it is also natural to ask whether the stress affects just public sector research or also characterizes private sector research. A recent report from the Task Force on the Health of Research (U.S. House 1992) provides strong evidence that the malaise is not limited to agriculture and characterizes at least the federal research system. That report cites an important paradox that has been associated with the performance of research in the past (see e.g. Markoff, 1990; Lederman, 1991; Council on Competitiveness, 1991; White, 1991). Over the past fifty years, the U.S. has supported a scientific research system which has been highly productive in generating new technologies and knowledge that have served the goal cited by Vannevar Bush to "insure our health, prosperity, and security as a nation in the modern world". Despite this apparent success, the U.S. paradoxically remains plagued by societal crises of grave magnitude and wide ranging character. The Task Force report concludes that the existence of this *general paradox of research* suggests that either the results of research are not being utilized, or that the research system is not producing the results needed to solve current challenges.

The report argues that the performance of the U.S. research system may be critically dependent on the structure of the research system including the linkage of research to explicit national goals, and the design of how research is accomplished. In particular, the report recommends that research must be directed at national goals, and that the conduct of research should ensure that optimal paths to attainment of those goals are taken, and that research performance must be evaluated objectively from a societal perspective, not by the research community itself. The report cautions that enhancing performance of the research system must not be interpreted as solely determined by levels of funding.

The chapters of this volume examine from various perspectives the existence and nature of this general paradox of research as it may currently affect agricultural research. The chapters add clarity to the view that the U.S. agricultural research system is currently under substantial challenge. Broadly, three types of challenges were identified that face the system: 1) changes in the structure and focus of demands placed on the system, 2) changes in the nature and extent of research opportunity (as defined by scientific opportunity, regulation, and economic incentives including intellectual property rights), and 3) changes in the level and nature of funding of agricultural research.

The objectives of this chapter are to summarize the salient observations of the forgoing chapters and to utilize them to consider the sources of stress on and nature of challenges faced by the U.S. agricultural research system. The chapter will consider these forces within the context of an economic framework that views the level of research effort and support for that effort as being determined through the interaction of societal demands for and supply of research solutions. From this reconsideration it will become clear that agricultural research is challenged not only by the general paradox of research discussed above, but also by two further paradoxes. First, there is the paradox of success, research has generated quantitative productivity gains that have generated excess supplies and reduced support for further research. The second paradox is interdependent with the paradox of success. The agricultural research system's past successes have apparently weakened its societal franchise to conduct research to solve the qualitative productivity issues the agricultural sector now faces. We label this the *paradox of relevance*. The chapter argues that together these three paradoxes are the source of the key strategic issues facing U.S. agriculture today.

Symptoms of Stress in Agricultural Research

The opening chapters of this volume cite a broad array of forces which are currently affecting the agricultural research system. In this section, these factors are summarized and available evidence is assessed concerning the extent and nature of any stress faced by the U.S. agricultural research system.

Woods notes that the success of the U.S. agricultural research system has been widely recognized, however, Woods argues that the dependence of that success on long-term investment in the system's essential scientific infrastructure and capacity has not been appreciated. According to Woods, weakening support for U.S. agricultural research is usually based on a number of rationales. First, current surpluses of agricultural commodities apparently relax social imperatives for research directed at output expansion. Woods notes that such surpluses may be interpreted as an indicator that the industry has reached maturity, characterized by limited potential for growth in output relative to opportunities for enhancing the value of output through market development and commercialization. Reduced fiscal capacity at the state and federal level to finance public research, declining rural population, and growth of nonagricultural constituents with special interests in the performance of agriculture that go beyond output levels are also noted by Woods.

Busch adds the general paradox of research to the forces that have softened the support for agricultural research; namely, the evolution of declining societal interest and confidence in science as a source of solutions. Busch notes this declining public confidence in science and its products has been observed across numerous disciplines and economic sectors. Busch notes that for agriculture this trend has left many with the view that public support for research is simply a form of sectoral subsidy paralleling farm programs that support farm prices and income.

Herrett notes these same factors, however, he also assesses the extent to which priorities of agricultural research have shifted in focus to issues which have emerged from new clientele groups. Importantly, Herrett notes that the priorities recognized by the Joint Council of Food and Agricultural Sciences (JCFAS 1992) no longer include increased productivity, while issues relevant to social impacts of agricultural activities now dominate this list of priorities. Herrett argues that despite the emergence of new priorities, the importance of the traditional focus on quantitative productivity must not be obscured. To do so, in Herrett's opinion, would overlook the long-run interests of society in enhancing quantitative productivity.

Despite the forces identified in the first section, the empirical evidence presented by Pray, Duvick, and Schweikhardt and Whims (S&W) suggests that public support for agricultural research through 1990 has continued to increase, though as Duvick notes at a declining rate of growth. Pray notes that in FY1991, real state funding declined by about one percent. Nonetheless, Pray notes that federal and private funding increases to state agricultural experiment stations (SAES) compensated for this decrease. Further, as Pray notes, in the public sector, basic research has increased, as has research aimed at quantitative productivity. Only during the past five years has quantitative productivity research declined slightly relative to environmentally related research. Pray indicates that the slowed growth in public research was accompanied by increased growth in recent years for private sector research, with private sector research now accounting for about 55 percent of all agricultural research expenditure in the U.S. Comparing these trends to those for other countries, Pray finds no basis for alarm with respect to potential loss in competitiveness of U.S. agriculture. In both the private and public sector, U.S. research expenditure growth remain dominant in the world.

Duvick notes these same trends though emphasizes that the emergence of societal priorities that go beyond quantitative productivity can be expected to threaten research funding in quantitative productivity and lead to a shift in focus to these new concerns. Duvick further suggests that increases in state funding for these issues are not likely and expects the SAES will necessarily turn to new sources for funding for such research.

Further appreciation of these trends can be gained by placing them in historical context. While funding levels continue to grow at a declining rate, Schweikhardt and Whims (S&W) note that this must be compared to the period of 1945 to 1970 during which federal funding of SAES system increased at about 10 percent annually. Since that time, growth in federal funding has declined, as has that of state funding, the share of federal funding has declined, and the share of state funding has remained roughly flat. Growth in total funding has been preserved through increases in private sector and nonagricultural funding of public sector research at the SAES. Accompanying these changes in funding, substantial shifts have occurred in the composition of funding among alternative mechanisms for allocating federal research funds (S&W). In particular, since the early 1980's *formula based* funds have declined from over 98% of CSRS funds to about 50%, a decline of about 3% per year in real terms since 1982. Replacing this funding mechanism have been *special* and *competitive grants* which have each grown to about 25% of CSRS funds. On a per scientist basis, special grants have grown at about 8% per year, while competitive grants have expanded by about 13% per annum. These trends have forced the SAES system to rely increasingly on state and other sources of funding for base support.

More recent data are not formally available, however, available evidence suggests that the process of declining support for agricultural research has accelerated in the past three years. A recent report by the Experiment Station Committee on Organization and Policy (ESCOP) "State Appropriation for SAES" dated April 4, 1993 presents budget data for FY1990-1993. The report finds that state appropriations increased about 1.4 percent from 1990 to 1993, however, after inflation (based on state/local CPIs) this represented a 6.6 percent real decrease in the research budget line. This trend was accompanied by a 6.4 percent decrease in the number of scientist positions supported by SAES funds. On a disaggregated basis, seventeen states were found to have experienced decreases in funding between 1990 and 1993, while 25 percent of the SAES directors indicated they expected slight to great decreases in state appropriations for 1994 and 1995. A recent study of funding of SAES through State Appropriations for FY1990 to FY1993 provides further detail to this image (Klonglan, 1993). Klonglan's findings indicate that the distribution of the states in which SAES have faced decreases in budget is skewed to those with the largest budgets. Regionally, his results indicate that the SAES in the northeast and south have suffered disproportionately. In general, in these regions, the SAES suffered greater decreases than the their universities as a whole.

These trends can be viewed as manifestations of the general process of weakening societal support for science as noted by the Task Force Report (U.S. House, 1992); however, they must also be considered as part of long term process that is specific to agriculture. The tension between societal interest in and valuation of basic versus applied research has accompanied considerations of funding of the Land Grant system since its inception (Marcus, 1986; Knoblach et al., 1962; True, 1937) and has re-emerging repeatedly (NRC, 1985; Winrock, 1982; NRC, 1972).

A Reconsideration of the Origins of Stress

Several conclusions can be drawn from the above summary. First, the demand for public agricultural research solutions originating from farmers has remained flat or declined over the past decades as private input industries have emerged, as they have increased their role in supplying research solutions, and as productivity problems have become increasingly farm specific. Second, the demand for research solutions to manage the externalities and qualitative aspects of agricultural production has expanded over the past decades as society has become increasingly aware of these agricultural outputs. Third, the public sector research capacity was increasingly criticized for failing to respond to these new demands for research, weakening its social franchise to conduct research to solve agriculturally related research problems. Lastly, the public sector research system has become cognizant of these new demands and has implemented approaches to address them.

These conclusions suggest that the level of stress on the U.S. agricultural research system is substantial. However, it is unclear from the combination of perspectives offered whether the root of the stress on the system lies in declining demand for agricultural research services or declining demand for such services as offered by a subset of or the whole of the institutions traditionally involved in agricultural research. In this section, I would like to consider this question within an economic framework. To do so, the symptoms of stress are organized according to general forces involved in motivating and rationalizing research activity, namely the demand for and supply of research solutions, constraints on private and public decisions to invest in agricultural research, and the evolution of alternative roles played by the private and public sectors.

Private vs. Public Sector Roles

To understand any shifts in private and public sector roles, the rationales for their roles in conducting research need to be distinguished (see e.g. Nelson, 1959; Arrow, 1962; Averch, 1985; Dasgupta, 1987; Stoneman and Vickers, 1988). Economists have long argued that this distinction properly rests on the appropriability and specificity of returns to research. Many innovations directly and specifically benefit the user. Innovations are appropriable if the innovator can control access to them both generally through prohibition of general use of the innovation and specifically through control of potential user access. When innovations are appropriable, users can be expected to be willing to pay for the benefit received through use. In this case, the innovation can be called a *private good* and the innovator may earn economic returns from marketing the innovation which are sufficient to finance its research and development. The role for the public sector in this type of research is limited and most often relies on existence of uncertainty in the payoff for research (Nelson, 1959). Alternatively, where the benefits of the use of an innovation are wide spread, and use of the innovation is not appropriable, a role of public sector research is strongly justified. In this case, economists classify the innovation as a *public good* (Nelson, 1959). Agricultural examples of this type of research innovation include planting practices, and disease or insect control tactics such as IPM. For this type of innovation, incentives for private research and development effort would be absent since farmers could adopt such innovations without paying for their use. Private market economics would fail to generate the research, despite possibly extensive societal benefits.

A further case exists for a public role in research even where the innovation may otherwise be a private good. In this case, the transactions and acquisition costs of the innovation may be so great as to preclude private sector dissemination of the innovation, implying also that no incentive would exist for research and development of the innovation. This situation has been recognized to have characterized the U.S. in the nineteenth century (Marcus, 1986) when transportation, communication, and institutional costs were very high. In this case, a strong rationale exists for a role for public sector provision of research and extension services. However, as these transactions costs decline with economic development, the rationale for public sector activity may be expected to disappear. Importantly, to ensure that private sector services emerge, it is critical that the public sector role evolve, catalyzing the private sector's participation.

Evolution of the Demand of Research

The demand for research solutions to improve the performance of the U.S. agricultural and food system has been characterized by at least three widely recognized phenomena over the past decade. As a symptom of the extent and nature of changing demands for research, vocal interest groups have emerged offering an extensive critique of the research system's performance and demanding a role in directing the agenda for research (for discussion see Busch, in this volume; Woods, in this volume; Lacy and Busch, 1986; for examples see: NRC, 1985; Kenny and Kloppenburg, 1983; Winrock, 1982; Hightower, 1978; NRC, 1972; Carson, 1962)). Secondly, on a wider basis, evidence suggests that society has lost some of its confidence in the ability of the agricultural research system to supply solutions and recommendations for public policy (Busch, Woods, Herrett, and Duvick, in this volume). Related to these first two phenomena is the third, the claim of a growing incongruence between societal demands for research solutions, expectations and goals of agricultural research system participants, and the goals and needs of policy decision makers (Pray, in this volume). These forces have led to increased demands for formalization and centralization of the agenda formation process guiding public sector research and, increased regulation of private sector research (Weaver, Townsend, and Huttner, each in this volume). Further, it is important to note that the perceived divergence between society's research goals and those of the traditional agricultural research system has led social interest groups to seek solutions from sources outside of that system (Busch). By implication, the social franchise of the traditional participants of the agricultural research system to conduct research to solve agricultural research problems identified as relevant by society has been weakened.

A further trend cited by the authors of the chapters of this volume suggests an additional factor in the evolution of the demand for agricultural research from traditional public sector participants. During the past several decades, the role of the private sector has expanded while that of the public sector has remained nearly flat (Pray, in this volume; Duvick, in this volume). The causes of this trend have not been formally investigated; however, one can speculate that several factors may have contributed. First, during the postwar period, institutional, communication, and transportation innovations have fundamentally altered the transactions costs which prohibited private sector provision of many research and development services. Second, shifts in intellectual property rights and the potential to protect innovations have altered incentives for private sector innovations in both plant and animal sciences (Lesser, in this volume; Duffey, in this volume). Finally, declining support for public sector research and development has augmented the demand for private sector services.

Evolution of the Supply of Research

On the supply side, the success of agricultural research in contributing to the quantitative productivity of resources used in agriculture is often cited (Fox et al., 1987; Evenson et al., 1979;

Evenson, 1978). Nonetheless, the appropriateness of that goal as well as the path taken for its achievement has been increasingly questioned as demand has evolved to include broader social interests (for discussion see: Busch, in this volume; for examples see Kenny and Kloppenburg, 1983; Winrock, 1982; Hightower, 1978). At the same time, the agricultural research system has been criticized as not having pursued new opportunities available through advances in basic science (e.g. NRC, 1985; NRC, 1972), and the need for increased effort in basic sciences has been emphasized (NRC, 1989). Importantly, advances in basic science have presented important opportunities for solution of both traditional quantitative productivity problems as well as of many of the new externality and qualitative productivity issues identified by new nonagricultural constituents such as water quality, impacts of agricultural chemical use, or food safety.

The decision to invest in agricultural research follows distinct protocols in the private sector as compared to the public sector. Trends in the levels of funding can be interpreted as reflecting a changing balance between demand and supply for research and the impact of that balance on research investment decisions. The effects of flat federal funding and declining state funding have been to soften the financial support for long-term research projects and to pose a possible threat to the long-term feasibility of particular research activities as infrastructure capacity and human resources are reduced (Holt, in this volume; and Woods, in this volume). While private funding of public research has increased (Pray, in this volume; Duvick, in this volume; and S&W, in this volume), private funding has typically had shorter horizons than public funding and has failed to substitute for reductions in base funding from public sector sources duvick, in this volume).

The Balance of Demand and Supply

These trends in demand and supply may be interpreted as resulting in increased competition for available private and public research budgets (Busch, in this volume; Woods, in this volume). This competition is heightened by the loss of political power of farm interest groups and the associated loss in their ability to direct research funds toward agricultural productivity issues (Busch, in this volume; Woods, in this volume). However, the declines in funding may also reflect an increasing failure of societal consensus to accept the rationale for public funding of agricultural research (Busch, in this volume). At the extreme, the argument has been made that public funding of agricultural research to enhance quantitative productivity must be interpreted as simply an alternative form of subsidy to an already highly subsidized sector of the economy (Doyle, 1985; Hightower, 1973). Consistent with this critique is the argument that the dramatic growth in agricultural productivity during the post-war period has typically outstripped growth in demand leading to persistent excess supplies. Within this context, the rationale for private or public funding of agricultural research has been argued to be weak. We label this the *paradox of success* of agricultural research.

In both the private and public sectors, a further development has been the diversion of limited research budgets. The evolution of societal demands for externality and qualitative productivity research has impacted both the private and public sector research budgets. In the private sector, regulatory requirements have forced significant competition between research and regulatory uses for available funds (Weaver, in this volume; Townsend, in this volume; Huttner, in this volume). In public sector, societal demands have been translated into directed funding administered through competitive grants. This has resulted in research effort and expenditure being diverted to compete for targeted funding from sources within the agricultural research system as well as from extramural sources (see discussions of Offutt, in this volume; Chubin, in this volume). As a result, in both private and public sectors, competition for limited research funds has increased. In the public sector, this trend of increased competition has led to increased pursuit of and reliance on special grant funds from public sources (Chubin, in this volume; S&W, in this volume).

Interpretation

When viewed from the perspective of the evolution of the agricultural sector in the United States, the shifts in the composition of demand for agricultural research seem almost natural. In the mid-nineteenth century, agriculture was highly decentralized and the role of private sector input and technology supply industries was minimal. Given the decentralization of farm production, significant information and transactions costs meant that innovative technologies did not diffuse quickly. The incentives for private sector initiatives in this regard were weak. Within that context, substantial benefits could be expected from public sector efforts to develop and diffuse technologies directly to farmers. Because these technologies were generally of value to most farmers they can be thought of as *generic technologies*.

This direct linkage of research and diffusion with the farm clientele meant that the value of research was apparent to a farm clientele with an high stake in continued access to the results of research. Not surprisingly, farm support for public funding of agricultural research was high. Given that the population was, in general, rural during those years, this support was effective in garnering public funding. However, as opportunities for diffusion of generic technologies diminished as they were increasingly adopted, and as information and transactions costs of diffusing input supplies and technologies declined, incentives appeared for a private sector role in these activities. As the agricultural input and technology industry emerged during the first half of the twentieth century, the industry increasingly became the source to which farmers turned for solutions. At the same time, the industry turned to the public research system for solutions and, where profitable, began to develop their own capacity to conduct and develop research. An implication of these changes was that the clientele of the U.S. agricultural research system began to evolve from being largely composed of farmers to being dominated by input and technology supply firms (Marcus, 1986; Schweikhardt and Bonnen, 1986). Simultaneously, the evolution of large scale farms with high financial stakes has increased the demand for and use of specialized consultants and contracted services. First, these sources provide a means of obtaining farm specific information that offer a basis of enhanced performance at the competitive margin. Second, because they offer proprietary information, they may be perceived by farmers as having a greater stake in the farm's performance than public sources. As this occurred the distance between the agricultural scientist and the farmer increased as the farmer became increasingly abstracted from the contribution of agricultural research.

The recent emergence of public interest groups with demands for specific types of research solutions can be viewed as a further step along an evolutionary path of the composition of the demands for agricultural research. As these demands for solution of the externalities and the qualitative implications of agricultural production emerged and expanded, the SAES were viewed as serving input and technology supply firms and were not viewed as a credible source for solutions to these emerging problems. Based on this logic, often expressed in criticism of the performance of the SAES, these interest groups turned to other research sources, and argued for a disenfranchisement the agricultural research system from its previous role as a predominant source of public sector research solutions for agriculture. Moreover, many of the emerging issues were ones for which other public agencies had existing research capacity, e.g. food safety, agriculturally originating pollution, nutritional quality (GAO, 1993). These forces left the traditional public sector research capacity with a possible weak indirect role in the solution of these newly emerging problems.

Historically, the charge of the U.S. agricultural research system has been to expand quantitative productivity and income of agriculture (Knoblach, et al., 1962; True, (1937). Through much of the postwar period these narrow sectoral goals were widely thought of as coincident with enhancing social welfare. For the private sector, such a focus is synonymous with enhancing profitability, a prerequisite for innovations to be purchased by farmers and integrated into their farm production practices. For the public sector, the charge of agricultural research emerged from the debates over the establishment and purpose of the SAES with the passage of the Hatch Act (Marcus, 1985). Of great concern was the extent to which the agenda for research would be set by farm needs or by agricultural scientists. The debate evolved toward an uneasy equilibrium in which agricultural research attempted to serve farmer's needs while solving the identified problems with scientific approaches (Rossiter, 1975; Scott, 1972). To this extent, the focus of the SAES has traditionally been client oriented. However, as the preceding section highlighted, the clientele has evolved from farm to a broader societal base and the political distance between both farm and broader social clientele has expanded. Thus, while the demand for agricultural research solutions to agricultural productivity problems (defined in the general sense to include externality and qualitative concerns) has increased substantially, and as the traditional elements of the agricultural research system has lost its franchise to conduct research, those demands are now being placed with and served by researchers and institutions that are outside of the traditional agricultural system.

Within this context, we can now reconsider several issues raised by the authors of chapters in this volume. First, what interpretation can be placed on declining or flat real funding of public sector agricultural research? Is this a result of political ineptness of the agricultural community, or more fundamentally a result of a natural evolution of the role of private sector research and marketing capacity? Second, what interpretation can be given to declining state funding?

On this first question, the authors of chapters in this volume have cited numerous indicators or symptoms of declining support for public sector research. However, this situation can not be claimed to be a new one facing agriculture (Marcus, 1986). Alternatively, it was argued in this chapter that this current cycle of diminished support might reflect the natural evolution of private sector capacity for delivery of research, as well as the evolution of the demands for research by the agricultural community and society, in general. Nonetheless, in view of the rapid expansion of demand for agricultural externality and qualitative productivity research, this declining support must be viewed as suggesting a substantial loss of society's franchise for conducting research by the traditional agricultural research system.

Emergence of a New Paradox

From this perspective, a new paradox can be identified that faces the traditional participants in the agricultural research system: *the paradox of relevance*. The traditional system has been viewed as successful in generating solutions to quantitative productivity problems. Society has increasingly viewed farm production as an open system, using a wide array of environmental inputs and producing a wide array of qualitative and externality outputs. This view has brought new constituencies to support research to solve these externality and qualitative productivity problems. The paradox of relevance is that given success in quantitative productivity research and growth in demand for externality and qualitative productivity research, society has turned away from the traditional agricultural research community for solutions to these new problems. Far from turning to the traditional participants, society has expressed the urgency of its demands by regulating private sector research to ensure achievement of social goals and by increasingly directing public sector research to serve societal demands by using competitive granting and public agenda processes (NRC, 1989; Offutt, in this volume). In combination with the paradox of success faced by agricultural research these dual paradoxes have placed a substantial challenge to the system to rationalize and finance both its traditional research on quantitative productivity and a possible role in the provision of solutions to externality and qualitative productivity issues.

References

Adams, James D. 1990. "Fundamental Stocks of Knowledge and Productivity Growth." *Journal of Political Economy*. Vol. 98, No. 4, pp. 673-702.

Arrow, K. 1962. Economic Welfare and the Allocation of Resources for Invention, in R. Nelson (ed.), The Rate and Direction of Inventive Activity (Princeton University Press, New Jersey).

Averch, H. 1985. A Strategic Analysis of Science and Technology Policy (John Hopkins Press, Baltimore).

Bloch, Erich. 1991. "Optimists, Skeptics, and Realists: Other Views of the Research 'Crisis'." In *Science and Technology Policy Yearbook*.

Busch, Lawrence, in this volume.

Busch, Lawrence and William B. Lacy. 1983. *Science, Agriculture, and the Politics of Research*. Westview Special Studies in Agriculture Science and Policy. Westview Press, Boulder, CO.

Busch, Lawrence and William B. Lacy. 1986. *The Agricultural Scientific Enterprise*. A System in Transition. Westview Special Studies in Agriculture Science and Policy. Westview Press, Boulder, CO.

Bush, V. 1945. Science: The Endless Frontier (National Science Foundation, Washington) (reprinted, 1960).

Carson, R. 1962. *Silent Spring*. Houghton Mifflin, NY.

Chubin, D. E., in this volume.

Chubin, D. E. 1986. "The Metaphor and its Measure." *BioScience*. Vol. 36, pp. 158-159.

Chubin, D. E. (February) 1991. "Progress, Culture, and the Cleavage of Science From Society." *Science, Technology, and Social Progress*. S. L. Goldman (ed.) (Bethlehem, PA: Lehigh University Press, 1989), pp. 177-195; and "Is Knowledge a Dangerous Thing?" *The Economist*, Vol. 318, pp. 21-22.

Congress of the United States, Office of Technology Assessment. 1991. *Federally Funded Research: Decisions for a Decade*. U.S. Government Printing Office. Washington, D.C.

Culliton, B. J. 1982. "The Academic-Industrial Complex." *Science*. Vol. 216, pp. 960-962.

Dasgupta, P. 1987. The Economic Theory of Technology Policy: An Introduction in: P. Dasgupta and P. Stoneman, *Economic Policy and Technological Performance* (Cambridge University Press, Cambridge).

David, P.; D. Mowery; and W. Steinmuller. 1988. The Economic Analysis of Payoffs from Basic Research -- An Examination of the Case of Particle Physics Research, *CEPR Publication No. 122*, Center for Economic Policy Research, Stanford University, California, 1988.

Dertouzos, M.; R. Lester; and R. Solow (eds.). 1989. *Made in America: Regaining the Productive Edge* (MIT Press, Cambridge, MA).

Doyle, J. 1985. *Altered Harvest*. New York: Viking.

Duffey, W., in this volume.

Duvick, D., in this volume.

Evenson, R. E. (December) 1983. "Intellectual Property Right and Agribusiness Research and Development: Implications for the Public Agricultural Research System." *American Journal of Agricultural Economics*. pp. 967-975.

Evenson, R. E. 1978. "A Century of Productivity Change in US Agriculture: An Analysis of the Role of Invention, Research and Extension." Discussion Paper No. 196. Economic Growth Center, Yale University, New Haven, CT.

Evenson, Robert E.; Waggoner, Paul E.; and Ruttan, Vernon W. 1979. "Economic Benefits from Research: An Example from Agriculture." *Science*. Vol. 205 pp. 1101-1107.

Experiment Station Committee on Organization and Policy (ESCOP) and Cooperative State Research Service. 1984. *Research 1984*. Washington, D.C.: U.S. Department of Agriculture.

Fox, G. 1985. *Optimal Public Investment in US Agricultural Research*. A Study of the Management of Technical Change. Ph.D. Dissertation. University of Minnesota: St. Paul.

Fox, G.; R. E. Evenson; and V. W. Ruttan. 1987. "Balancing Basic and Applied Science: The Case of Agricultural Research." *Bioscience*. Vol. 37, pp. 507-509.

GAO. 1993. *Revitalizing USDA, A Challenge for the 21st Century*. United States General Accounting Office. Washington, D.C.

Garfield, Eugene. (November) 1990. "Assessing the Benefits of Science in Terms of Dollars and Sense." *The Scientist*. Vol. 4, No. 22, p. 14.

General Accounting Office. 1993. *Federal Agricultural Research Funding: Issues and Concerns*. Washington, D.C.: General Accounting Office.

Goodwin, Irwin. 1989. "Universities Reach into Pork Barrel with Help from Friends in Congress." *Physics Today*. pp. 43-45.

Hadwiger, D. F. 1982. *The Politics of Agricultural Research*. Lincoln: University of Nebraska Press.

Hague, D. (ed.). 1990. The Management of Science (Macmillan).

Herrett, R., in the volume.

Hightower, J. 1972. *Hard Tomatoes, Hard Times: The Failure of the Land-Grant College Complex*. Washington, D.C.: Agribusiness Accountability Project.

Hightower, J. 1973. *Hard Tomatoes, Hard Times*. Cambridge, MA: Schenckman.

Holt, D., in this volume.

Hullar, Theodore L. 1986. "Adaptations in the Agricultural Research System That Could Improve Our Ability to Capitalize on New Opportunities." *Agricultural Science Policy in Transition*. pp. 203-211.

Huttner, S., in this volume.

Joint Council on Food and Agricultural Sciences. 1994. "Priorities for Research, Extension and Higher Education: A Report to the Secretary of Agriculture."

Kenny, D. 1985. "Government Policies and the Cost of Doing Research." *Science*. 227, pp. 480-84.

Kenny, M. and Kloppenburg, J., Jr. 1983. "The American Agricultural Research System: An Obsolete Structure?" *Agricultural Administration*. Vol. 14, pp. 1-10.

Knoblauch, H. C.; Law, E. M.; and Meyer, W. P. 1962. *State Agricultural Experiment Stations, A History of Research Policy and Procedure*. Miscellaneous Publication, 904. Washington, D.C.: USDA.

Lederman, Leon M. 1991. *Science: The End of the Frontier?* President-Elect, to the Board of Directors of the American Association for the Advancement of Science.

Lesser, William, in this volume.

Lipman-Blumen, J. 1984. *The Paradox of Success: The Impact on Priority Setting in Agricultural Research and Extension*. Office of the Assistant Secretary. Science and Education, United States Department of Agriculture, Washington, D.C.

Marcus, Alan I. 1985. *Agricultural Science and the Quest for Legitimacy: Farmers, Agricultural Colleges, and Experiment Stations*. Ames, Iowa: Iowa State University Press.

Marcus, Alan I. 1986. From State Chemistry to State Science: The Transformation of the Idea of the Agricultural Experiment Station. *The Agricultural Scientific Enterprise*. Lawrence Busch and William B. Lacy, (eds.). Westview Press: Boulder and London.

Markoff, John. (January) 1990. "A Corporate Lag in Research Funds is Causing Worry." *New York Times*. p. 1.

Marshall, Eliot and Hamilton, David P. (November) 1990. "A Glut of Academic Pork." *Science*. Vol. 250, pp. 1072-1073.

Mayer, A., and Mayer, J. 1974. "Agriculture: The Island Empire." *Daedalus*. Vol. 103, pp. 83-95.

McCalla, A. 1978. The Politics of the Agricultural Research Establishment." In D. F. Hadwiger and W. P. Browne (eds.), *The New Politics of Foods*. Lexington, Massachusetts: Lexington Books.

Mowery, D. 1983. Economic Theory and Government Technology Policy. *Policy Sciences*. Vol. 16, pp. 27-43.

Mowery, D. and N. Rosenberg. 1989. *Technology and the Pursuit of Economic Growth*. Cambridge: Cambridge University Press.

NASULGC Board on Agriculture. 1993. *The U.S. Agriculture, Food, and Environmental System*. Averting A Crisis by Building for Future Success: An Imperative for the 1990s. Washington, D.C.

National Science Foundation. 1990. *Federal Funds for Research and Development -- Detailed Historical Tables: Fiscal Years 1955-1990*. Washington, D.C.

National Research Council. 1972. Report of the Committee on Research Advisory to the US Department of Agriculture. Committee Advisory to the USDA. Washington, D.C.: National Academy of Science.

National Research Council. 1985. *New Directions for Bioscience Research in Agriculture: High Reward Opportunities*. Committee on Bioscience Research in Agriculture. Washington, D.C.: National Academy Press.

National Research Council. 1989. *Research on the Management of Technology: Unleashing the Hidden Competitive Advantage*. Washington, D.C.: National Academy Press.

National Research Council. 1989. *Investing in Research. A Proposal to Strengthen the Agricultural System*. Washington, D.C.: National Academy Press.

Nelson, R. 1959. The Simple Economics of Basic Scientific Research. *Journal of Political Economy*. Vol. 67, pp. 297-306.

Nelson, R. 1987. *Understanding Technical Change as an Evolutionary Process*. North-Holland: Amsterdam.

Nelson, R. and R. Levin. 1986. The Influence of Science University Research and Technical Societies on Industrial R&D and Technical Advance. Policy Discussion Paper Series Number 3, Research Program on Technology Change. Newhaven, CN: Yale University.

Office of Technology Assessment. 1981. An Assessment of the United States Food and Agricultural Research System. OTA-F-272. Washington, D.C.: U.S. Government Printing Office.

Offutt, Susan, in this volume.

Pavitt, Keith. 1987. The Objectives of Technology Policy. *Science and Public Policy*. Vol. 14, pp. 182-188.

Pavitt, Keith. 1991. "What Makes Basic Research Economically Useful?" *Research Policy*. Vol. 20, pp. 109-119.

Powell, Orey S. 1989. "Universities Reach Into Pork Barrel with Help from Friends in Congress." *Physics Today*. Vol. 42, No. 4. pp. 43-45.

Pray, C., in this volume.

Rasmussen, W. D. 1982. "The Mechanization of Agriculture." **Scientific American**, Vol. 247, No. 3, pp. 77-89.

Rossiter, M. W. 1975. *The Emergence of Agricultural Science. Justus Liebig and the Americans, 1940-1980.* New Haven: Yale University Press.

Rosenberg, N. 1990. "Why do Firms do Basic Research (with Their Own Money)?" *Research Policy*.

Rhodes, James. 1986. *Agricultural Science Policy in Transition*. Agricultural Research Institute (ARI). Bethesda, MD.

Rowell, John M. (May) 1992. "Condensed Matter Physics in a Market Economy." *Physics Today*. pp. 40-47.

Schweikhardt, David B. and James T. Bonnen. Policy Conflicts in Agricultural Research: Historical Perspective and Today's Challenges. The Agricultural Scientific Enterprise. In Lawrence Busch and William B. Lacy, (eds.). Westview Press, Boulder and London.

Schweikhardt, David and John Whims, in this volume.

Science for Agriculture. 1982. Report of a Workshop on Critical Issues in American Agricultural Research. Winrock International Conference Center, Morrilton, Arkansas.

Scott, R. V. *The Reluctant Farmer. The Rise of Agricultural Extension to 1914.* Urbana: University of Illinois Press.

Stoneman, P. 1987. *The Economic Analysis of Technology Policy*. Oxford: University Press.

Stoneman, P. and J. Vickers. The Assessment: The Economics of Technology Policy. *Oxford Review of Economic Policy.*

Townsend, Rod, in this volume.

True, A. C. 1937. *A History of Agricultural Experimentation and Research in the United States, 1607-1925.* Washington, D.C.: Government Printing Office.

U. S. Government Printing Office. 1992. *Report of the Task Force on the Health of Research.* Chairman's Report to the Committee on Science, Space, and Technology U.S. House of Representatives. Washington, D.C.

U. S. Congress, Office of Technology Assessment. 1991. *Federally Funded Research: Decisions for a Decade.* OTA-SET-490 Washington, D.C.: U.S. Government Printing Office.

White, Robert M. (Spring) 1991. "Too Many Researchers, Too Few Dollars." *Issues in Science & Technology.* Vol. 7, pp. 35-37.

Weaver, R., in this volume.

Weaver, Robert D. 1986. *Federal Research and Development and U.S. Agriculture: An Assessment of Role and Productivity Effects.* Paper presented at symposium: The Federal Role in Research and Development, November 1985, Washington, D.C., sponsored by the Committed on Science, Engineering and Public Policy, National Academy of Sciences.

Woods, Walter, in this volume.

About the Contributors

Lawrence Busch is professor of sociology at Michigan State University. He has been studying the organization and structure of domestic and international agricultural research for the past 15 years. He is author or co-author of six books and numerous articles on the subject, the most recent of which is *Plants, Power, and Profit: Social, Economic, and Ethical Consequences of the New Biotechnologies*. His current research focuses on the conservation of biological diversity in crop plants.

Daryl E. Chubin is Senior Associate in the Science, Education, and Transportation Program, Office of Technology Assessment, U.S. Congress. Before joining OTA in 1986 as Senior Analyst, he taught for 14 years at Southern Illinois University (Edwardsville), Cornell University, the University of Pennsylvania, and Georgia Institute of Technology. He was Professor in the School of Social Sciences at Georgia Tech and former director of its M.S. Program in Technology and Science Policy. He is a 1968 graduate of Miami University (Ohio), earning A.M. and Ph.D. (1973) degrees at Loyola University (Chicago).

Dr. Chubin's research has centered on the social and political dimensions of science and technology: science policy (especially as related to research misconduct and peer review), public understanding of science, interdisciplinary teamwork, and education and human resources. He has published numerous articles, chapters, and commentaries, and five books, including *Science Off the Pedestal: Social Perspectives on Science and Technology (coedited, Wadsworth, 1989) and Peerless Science; Peer Review and U.S. Science Policy* (coauthored, SUNY Press, 1990). Chubin was Project Director for OTA's May 1991 report, *Federally Funded Research; Decisions for a Decade*, and the June 1988 report, *Educating Scientists and Engineers: Grade School* to Grad School. He is a contributing editor to *BioScience* and an advisory editor to four other journals. In 1990 he was elected a Fellow of the American Association for the Advancement of Science.

William H. Duffey, General Patent Counsel with the Monsanto Company in St. Louis, Missouri, is an expert in the field of worldwide patent issues in biotechnology. He has served, twice in 1986 and again in 1987, as the legal advisor and biotech consultant to the United States Delegation to the World Intellectual Property Organization

(WIPO) in Geneva. WIPO is an agency of the United Nations having 100 member countries.

Mr. Duffey is a member of the American Bar Association, the American Intellectual Property Law Association, the Licensing Executives Society, the Association of Corporate Patent Counsel, and the Intellectual Property Owners, Inc. He is also a member of the Bars of Missouri, the U.S. Patent & Trademark Office and U.S. Court of Appeals for the Federal Circuit (CAFC).

William Duffey was born and raised in Indianapolis, Indiana, receiving his Bachelor of Science at the University of Notre Dame in 1949. From there he went on to the Indiana University Law School for his Doctor of Jurisprudence in 1960.

Donald N. Duvick is coordinator of corn breeding at Pioneer Hi-Bred International, Inc. Donald N. Duvick was born December 18, 1924 at Sandwich, Illinois. He received a B.S. degree in agronomy from the University of Illinois in 1948 and a Ph.D. in botany from Washington University in 1951. From 1951 to 1990, Duvick was employed by Pioneer Hi-Bred International, Inc. successively as Corn Breeder Coordinator of Corn Breeding, Director of Corn Breeding, Director of Plant Breeding, Vice-President/Research and Senior Vice-president/Research. His responsibilities included coordination of his company's plant breeding activities in corn, sorghum, wheat, soybeans, alfalfa and sunflowers. He was also a member of the Corporate Executive Committee and of the Corporation's Board of Directors. Following retirement from Pioneer Hi-Bred International, Inc., he has accepted an appointment as Affiliate Professor, Department of Agronomy, Iowa State University, Ames, Iowa.

Research contributions over the past 40 years have been in several fields, including the developmental cytology and biochemistry of starch and protein components of the maize endosperm, the genetics and practical applications of cytoplasmic male sterility in maize, elucidation of changes in productivity of commercial maize hybrids since the advent of hybrid maize, and genetic diversity as affected by plant breeding. He has published pioneering, frequently cited papers in each of these fields.

Duvick has published in numerous scientific publications and is a frequent reviewer for genetic and plant breeding journals. He lectures frequently at seminars and symposia in the U.S. and abroad, and serves on committees for the National Resources Council and the Office of Technology Assessment. He is President of the American Society of Agronomy. He has served in various capacities in the Crop Science Society of America, the American Seed Trade Association, the National Council of Commercial Plant Breeders, the National Plant Genetic Resources Board, the American Society of Agronomy, The Nature Conservancy, and the American Association for the Advancement of Science. Dr. Duvick served as assistant editor of the journal Plant Physiology in 1977 and 1978. He was a member of the organizing committee for the first International Crop Science Congress. He is a Fellow of the American Society of Agronomy, the Crop Science Society of America, the American Association for the Advancement of Science and is a Distinguished Fellow of the Iowa Academy of Science.

Richard A. Herrett is president of EnvirAg Associates, a consulting firm that specializes in agricultural and environmental issues. Clients of EnvirAg Associates include both large and small corporations as well as government regulatory agencies that have interests in pending legislation and existing regulations. Prior to establishing EnvirAg, Dr. Herrett was Government Relations Scientific Liaison for ICI Americas. In that position, he was responsible for representing ICI's broad range of technical interests including biotechnology, agriculture and the environment before the appropriate legislative and regulatory bodies. Previously, Dr. Herrett was Director of Research and Development for the Agricultural Division of ICI, a position he held since 1975. He joined ICI Americas in 1970 after 11 years as a Senior Scientist in the life sciences with Union Carbide and the Boyce Thompson Institute. A 1954 graduate of Rutgers University, Dr. Herrett holds a Master's Degree in agronomy and a Ph.D. in plant biochemistry from the University of Minnesota.

Donald Holt received his B.S. and M.S. degrees from the University of Illinois in 1954 and 1956, respectively. He operated his home farm and another farm in northern Illinois from 1956 to 1963. He undertook further graduate training at Purdue University, receiving his Ph.D. in 1967. He became a member of the Purdue faculty at that time and remained there until becoming Head of the University of Illinois Agronomy Department in May, 1982. In October 1983, he became Director of the Illinois Agricultural Experiment Station and Associate Dean of the University of Illinois College of Agriculture. Dr. Holt's research interests include environmental physiology of crops, agricultural systems analysis, computer simulation of agricultural systems, and forage crop management.

Wallace E. Huffman is Professor of Economics and Agricultural Economics, Iowa State University, Ames, IA 50011. His research areas are human capital, labor supply, agricultural economic growth and productivity, science for agriculture, and econometrica. He has a large number of publications in these areas and is co-author of a book entitled *Science for agriculture: A Long Term Perspective* (Iowa State University Press, 1993). This book provides evidence about the evolution, development, impacts, and funding of the U.S. public and private agricultural research and extension systems and an assessment of public policies to foster greater future productivity of the U.S. R&D system.

Susanne L. Huttner is Director of the University of California's Systemwide Biotechnology Research and Education Program, the state's initiative to promote development and understanding of biotechnology. Since its inception in 1985, the Program's competitive grants have supported new ventures in research and training in science and engineering, ranging from agriculture to biomedicine to marine sciences. Another innovative Program initiative is building a scholarly base of information on social issues related to biotechnology by supporting research and training in the social sciences and humanities. The Program monitors federal and state policymaking and provides scientific analysis and guidance to assist legislators and regulatory agencies. Dr. Huttner is the University's representative for the California Interagency Task Force on Biotechnology. The Program serves as a statewide clearing house on biotechnology and supports public education through training programs for teachers, farm advisors, home economists, and local government officials.

Dr. Huttner serves on the executive board of the Institute for Science in Society and on the States' Steering Committee on Biotechnology Oversight. Her research has focused on the molecular and cellular factors involved in the developing mammalian nervous system, combining molecular biology and electrophysiological techniques. She earned a Ph.D. in Neuroscience from UCLA and a Bachelor of Science degree from the University of California at Berkeley.

William B. Lacy is professor of rural sociology, Assistant Dean for Research and Assistant Director of the Pennsylvania Agricultural Experiment Station at The Pennsylvania State University. Dr. Lacy received his B.S. (1964) from Cornell University, his M.A. (1965) in administration in higher education from Colgate University, and his M.A. (1972) and Ph.D. (1975) in sociology from the University of Michigan. Prior to his arrival at The Pennsylvania State University, he spent fifteen years at the University of Kentucky where he was a Professor of Sociology, Director of the Food, Environment, Agriculture and Society in Transition program (W. K. Kellogg Foundation funded) and Co-chair of the Committee for Agricultural Research Policy.

Dr. Lacy has authored/co-authored over fifty journal articles and book chapters on education, science policy, agricultural research and biotechnology, and co-authored/co-edited five books: *Science, Agriculture and the Politics of Research (1983); Food Security in the United States (1984); The Agricultural Scientific Enterprise: A System in Transition (1986); Biotechnology and Agricultural Cooperatives (1983), and a recently published book entitled, Plants, Power, and Profit: Social, Economic and Ethical Consequences of the New Biotechnologies (1991).*

He has been a consultant for the U.S. Agency for International Development, the Congressional Office of Technology Assessment, U.S. General Accounting Office, and EMBRAPA (Brazil); a member of the Board of the Council for Agriculture, Science and Technology; and recipient of grants from the National Science Foundation, U.S. Agency for International Development, and the Ford Foundation. Dr. Lacy is a fellow of the American Association for the Advancement of Science and the recipient of the 1990 award for Excellence in Research from the Rural Sociological Society.

W. Lesser is Professor of Marketing, Department of Agricultural Economics at Cornell University. Areas of research and teaching specialization include the implications of patents for agriculture, international harmonization of intellectual property law, implication of biotechnology for the structure of farming, market structure and public policy, export marketing. Dr. Lesser earned his Ph.D. in Agricultural Economics at the University of Wisconsin in 1978 and has been at Cornell since that time.

Paul F. O'Connell serves as Deputy Administrator -- Cooperative State Research Service, U.S. Department of Agriculture.

He provides leadership for a national program whose objective is to find alternative opportunities for U. S. farmers. Major program areas are aquaculture, industrial uses, small-scale farming, and low-input/sustainable agriculture.

Prior to this position, Dr. O'Connell served for 3 years as Special Assistant to Orville G. Bentley, Assistant Secretary for Science and Education. Earlier professional experience involved 14 years with the Forest Service; 2 years with Economic Research Service; and 3 years in the private sector. With these organizations Dr. O'Connell served as Assistant Director, Project Leader, and Economic Analyst in Wisconsin, Arizona, and Minnesota.

Dr. O'Connell earned his B.S. degree from the University of Minnesota, majoring in biochemistry and dairy science. He obtained two master degrees -- one from Colorado State University and one from the Wharton School, University of Pennsylvania. These were in Agriculture Economics and Business, respectively. Ph.D. degree in Forestry was earned from the University of Wisconsin. He has received the USD Superior Service award, several cash awards and certificate of merits. He is author or co-author of several publications in science and education policy, natural resource issues and commercialization of promising technologies.

Susan Offutt is Executive Director in the National Research Council, National Academy of Sciences. Dr. Offutt joined the Board on Agriculture as its Executive Director in February 1992. The Board is a major program unit of the National Research Council, the principal operating agency of the National Academy of Sciences. Until early 1992, Dr. Offutt was Chief of the Agriculture Branch at the Office of Management and Budget in the Executive Office of the President. Before a tour at the U.S. Department of Agriculture's Economic Research Service in 1987, Dr. Offutt served as an assistant professor on the faculty of the College of Agriculture at the University of Illinois, Urbana-Champaign. Cornell University awarded her Ph.D. in agricultural economics in 1982.

Richard K. Perrin is Professor of Agricultural and Resource Economics at North Carolina State University, currently on leave at Iowa State University. He has also served as an economist at CIMMYT and in various leadership positions at NCSU and in professional associations. His research interests include the effects of intellectual property rights on innovation in agriculture, the contribution of genetic improvements to total agricultural productivity gains, the role of price and other policies in determining productivity gains, and in methods for ex-ante economic assessment of experimental technologies.

Carl E. Pray is Visiting Fellow, Economic Growth Center at Yale University and an Associate Professor of Agricultural Economics at Cook College, Rutgers University. He earned his B.A. at Carleton College in 1969 and a Ph.D. in Economic History at the University of Pennsylvania in 1978. From 1980 - 1986, he served as a research associate at the Department of Agricultural and Applied Economics, the University of Minnesota. He recently co-authored *Research and Productivity in Asian Agriculture*, published by Cornell Press in 1991.

Katherine (Kitty) Reichelderfer is currently Associate Administrator of the USDA's Economic Research Service, the agency responsible for providing social science information, research, and policy analysis to improve the performance of agriculture and rural America. Her previous experience includes several years as a Senior Fellow and Director of the Agriculture, Environment and Food Safety Program for the National Center for Food and Agricultural Policy at Resources for the Future.

Dr. Reichelderfer's research has focused on the economics of pest management, agricultural resource policy, and the interface of agricultural production and environmental quality, is widely published in a variety of journals, books, and popular outlets. Her M.S. and Ph.D. degrees in agricultural and resource economics were earned from the University of Maryland, where she earlier received a B.S. in biological sciences.

David B. Schweikhardt is an Assistant Professor in the Department of Agricultural Economics at Michigan State University. He received a Bachelor of Science degree in agricultural economics from Purdue University in 1980, a Master of Science in agricultural economics from Michigan State University in 1983, and a Ph.D. from Michigan State in 1989. From 1988 to 1992, he was an Assistant Professor of agricultural economics at Mississippi State University. His teaching responsibilities include courses in agricultural policy, international trade and research methodology and philosophy. His past research has examined the problems of financing agricultural research and extension activities in a federal system of government. His recent research included an examination of the impact of a North American Free Trade Agreement on the U.S. cotton industry. He currently holds a teaching-research-extension appointment in agricultural policy and trade.

Wolfgang E. Siebeck is an attorney. He holds degrees from Hamburg (Law) and Oxford (economics). He began his professional career in the Federal economic Ministry in Bonn, Germany, on development assistance to Africa. In 1969 he joined the World Bank where - at different stages - he was division chief in charge of lending strategies and programs to countries in North Africa and the Middle East, head of the Bank's Resident Mission in Pakistan, cofinancing manager for South Asia, and representative to the United Nations and the GATT (General Agreement on Tariffs and Trade) in Geneva, Switzerland. He retired from the World Bank in 1991 and now works in Washington, D.C. as a consultant to the Secretariat of, and research centers in, the Consultative Group on International Agricultural Research (CGIAR).

Rod Townsend is Regulatory Affairs Director at the Pioneer Hi-Bred International, Inc., plant breeding division. Dr. Townsend gained his B.S. degree in Microbiology from the University of Surrey, England, in 1972, and subsequently joined the John Innes Institute, Norwich, were he conducted basic research on plant pathogens for the Agriculture and Food Research Council. In 1975 he was awarded his Ph.D. in Plant Pathology from the University of East Anglia. He has published widely on the cellular biology of bacterial plant pathogens and molecular biology of plant viruses.

Townsend joined Pioneer in 1986 as a plant virologist. He was named to the position of Regulatory Affairs manager for Biotechnology in 1988 and was made Worldwide Director of Regulatory Affairs for the Plant Breeding Division in 1990. He is responsible for gaining regulatory approval for new products derived through biotechnology. He is a regular participant at U.S. and international meetings on biotechnology regulation and serves on a number of industry committees dealing with regulatory issues.

Robert D. Weaver is an Associate Professor of Agricultural Economics at Pennsylvania State University. His current research has focused on microeconomic and econometric analysis of public policy targeted at affecting the performance of agricultural production and marketing systems. Recent studies have considered agricultural impacts on water quality, agricultural policy and regulatory impacts on productivity, and the contribution of research to productivity, profitability and competitiveness of U.S. agriculture.

In 1977, he was awarded a Ph.D. by the University of Wisconsin in both Economics and Agricultural Economics with major fields in Public Finance and Agricultural Marketing, respectively. Since that time he has been a faculty member in the Department of Agricultural Economics at The Pennsylvania State University. During sabbatical, Dr. Weaver served as Senior Project Officer at the National Academy of Sciences, National Research Council, Board on Agriculture. Throughout his career, Dr. Weaver has served as a consultant to public agencies and the private sector including various U.S. House Agriculture subcommittees, the Council of Environmental Quality, the Office of Technology Assessment, the World Bank, and the U.S. Agency for International Development. Dr. Weaver also lectures annually as a visiting professor at the University of Clarmont-Ferrand, Centre d'Economie et Recerche de Developpment International, in France.

John F. Whims is a Graduate Research Assistant in the Department of Agricultural Economics at Michigan State University. He received a Bachelor of Science Degree in Business Administration from Greenville College (Illinois) in 1982 and a Master of Business Administration from Central Michigan University in 1984. He will receive a Ph.D. in Agricultural Economics from Michigan State University in 1993. He has consulted on strategic planning projects on the future of Michigan agriculture and on the political and economic environment affecting Michigan State University. He has also consulted with the Michigan Department of Corrections and the Senate Fiscal Agency. His doctoral research will examine policy issues in Michigan agriculture.

Walter R. Woods is currently Assistant Administrator of Regional Research, Cooperative States Research, USDA. He has served in this position since February 1993. Prior assignments were Dean of Agriculture, Director of Agricultural Experiment Station and Cooperative Extension Service at Kansas State University from 1985 to 1992. He served as Head of Department of Animal Sciences at Purdue

University from 1971 to 1985. Prior faculty assignments were at Iowa State University and University of Nebraska.

Brian Wright is a Professor of Agricultural and Resource Economics at the University of California at Berkeley. Brian is from Australia where his family had a sheep station in Southern New South Wales. He has a Bachelor of Agricultural Economics from the University of New England and received a Frank Knox Fellowship to do his Ph.D. in Economics at Harvard. He then taught in the Economics Department at Yale University. His research interests include economics of price supports, commodity markets and storage, as well as research and development policy.

David Zilberman is Professor of Agricultural and Resource Economics at U.C. Berkeley. He received his undergraduate degree in Economics and Statistics from Tel Aviv University in 1971. He graduated from U.C. Berkeley with a Ph.D. in Agricultural and Resource Economics in 1979. His dissertation addressed dairy waste management in Southern California.

Dr. Zilberman's recent work focuses on environmental and resource problems in agriculture -- in particular, water and pesticide problems. He is currently involved in studies for the EPA on water management in the West and the role of quality in marketing considerations for pesticide use choices. He has consulted to the EPA, USDA, World Bank, California State Department of Agriculture, CAL EPA, Westlands Water District, and the Environmental Defense Fund.

Abstracts

Chapter 1: The U.S. Agricultural Research System Under Challenge

Robert D. Weaver

The 1980's brought substantial changes in the focus and scope of agricultural research demanded by society. At the same time, the scientific opportunity with potential relevance for agricultural applications expanded dramatically. Together these fundamental changes demand and supply resulted in challenges to the U.S. agricultural research system to adapt. The expansion of the scope of demands placed on the agricultural research system was enormous. In one dimension, these demands expanded from those for new farm production technology to encompass the full realm of agricultural production and food processing all the way to the consumer's dinner table. In another dimension, demands for agricultural research expanded from those for enhancement of farm productivity and profitability to improving the *external* impacts of agricultural and food system production and marketing activities.

The expansion of the scope of demands on the agricultural research system (vertically through the food production and marketing system, as well as for the consideration of externalities) has far exceeded the expansion of the supply of agricultural research. The result can be viewed as a substantial excess demand for the services of the traditionally farm based agricultural research system. Both the scale and scope of this excess demand has challenged the U.S. agricultural research system to look beyond tradition and the *status quo* to find new means for response.

Chapter 2: U.S. Agriculture Research: An Assessment of the System

Walter R. Woods

Agricultural research has made significant contributions to U.S. and world agriculture in improving efficiency of production, assurance of a high-quality and safe food supply, addressing environmental issues, and conservation of natural resources. Research in support of U.S. agriculture has many forces shaping its future. These forces include loss of relative purchasing power, world competition in agriculture, other countries investing significantly in agricultural research, environmental issues related to agriculture and a loss of the focus on the importance of the role for the land grant university in agricultural research. The present strength of U.S. Agriculture is due in large part to a publicly supported research program. The opportunity for the U.S. for continuing a pre-eminent role requires that the public make investments in agriculture research. These past investments, both publicly and privately, have resulted not only in a wholesome and safe food supply, but one at a very competitive price. Many issues will shape the future of agricultural research and unless positive intervention is brought into the publicly supported research programs, the U.S. will experience a continued erosion of capacity and capability to the point that our competitiveness and profitability advantages in agriculture will surely be damaged.

Chapter 3: Agricultural Research in a Time of Change
Lawrence Busch

One hundred years ago, the noted soil scientist, E. W. Hilgard argued that the closing of the frontier would permanently transform agricultural research in the United States. Today, we stand at a similar juncture. Farmers are losing their special status in the eyes of the public. Environmental and consumer issues are eclipsing the more traditional concerns with production and productivity. Science, too, is changing as the post-harvest components of agriculture take on great significance in both public and private agricultural research.

Among the major areas of change are: (1) Increasing population without increasing effective demand for food, (2) greater concern for the environmental consequences of all aspects of agriculture, (3) global change which may change the cropping patterns, variability, and locations of particular crops worldwide, (4) public discontent with science and demands for specific definable benefits from scientific research, (5) growing consumer concerns about what they eat and how it is produced, (6) new patterns of world trade that favor regional trading blocs over nation-states, (7) pressure to eliminate farm programs, (8) potential new competitors in Eastern Europe, and (9) growing national debt forcing down the public agricultural research budget.

As a result agricultural research will have to become more responsive to a broader range of actors including consumers, environmentalists, backyard gardeners, rural bankers, rural electric cooperatives and agribusiness. Moreover, it will have to do this without losing its traditional farm clientele who are currently the only group that is willing and able to support public research at the state and national levels. Should the public sector fail in reorienting itself, it is likely that private research will be significantly handicapped as well.

Thus, the challenge of the next decades will be to broaden the agenda for agricultural research while increasing the base of financial support and satisfying the demands of an increasingly sophisticated public both here and abroad. Meeting that challenge will require all the skill and energy that researchers and research administrators can muster.

Chapter 4: Agricultural Research -- The Role of Change
Richard A. Herrett

The remarkable increases in agricultural productivity especially those of the most recent 5 decades were a direct consequences of the introduction of new technologies and the subsequent changes such as plant breeding resulting in hybridization; synthetic chemistry resulting in pest control and fertilizer techniques; engineering resulting in improvements in water management and mechanization. These technologies have existed for some time and are now approaching their upper limits. Unless new technologies such as biotechnology are developed and introduced into agricultural practices, the ability to incrementally increase agricultural productivity will diminish at a critical time when the world's population is expected to double in the next 5 decades.

There have been changes in the targets for agricultural research as illustrated by the comparison of the priorities determined by the Joint Council on Food and Agricultural Sciences. Such a comparison suggests a dramatic dilution of research efforts directed toward the discovery of new technologies leading to increased productivity. While certain of these additional priorities include strategies related to agricultural production such as sustainability and integrated pest management, other additional priorities are placed on research such as non-food uses of agricultural products and the sociological needs of rural America. These are unquestionably important priorities, but given the long lead times and the incredibly large costs required to introduce basic new technologies, one must not lose sight of the importance of researching and developing new technologies directed toward increasing productivity especially when the world's population is about to increase dramatically and at a time when the urban and environmental demands on land use are increasing at an accelerated pace.

Chapter 5: Trends in Food and Agricultural R&D: Signs of Declining Competitiveness?
Carl E. Pray

Growth in real agricultural research expenditure 1970 to 1990 by both the public and private sector has been fairly rapid. R&D expenditures by input and food industries continued to grow rapidly in the 1980s in contrast to private R&D growth in the seed, veterinary pharmaceutical and agricultural chemical industries where biotechnology is creating new opportunities. Public sector research expenditure also is growing. During the 1980s most of this due to increased state in funds but in 1990 and 1991 increases in federal funds have made up for the levelling off of state funding.

Relative to trends in public research expenditure most countries the U.S. still appears to be doing rather well. Most countries in Europe, with the exception of France and Spain, reduced public funding of agricultural research. Argentina and Brazil have drastically cut public spending on agricultural R&D.

Chapter 6: Funding Agricultural Research: An Assessment of Current Innovation
Donald N. Duvick

Agricultural Research funding, while not eroding drastically, is hardly keeping up with inflation in total. More importantly, sources and goals of funding are shifting, from public to private, from production to sustainability, from farming only to rural development, etc.

Chapter 7: Trends and Issues in Agricultural Research Funding at the State Agricultural Experiment Station
David B. Schweikhardt and John F. Whims

This paper examines trends in agricultural research funding at the State Agricultural Experiment Stations from 1970 to 1990. Emphasis is placed on examining (a) the changing composition of funding sources at the experiment stations, (b) changes in the mechanisms used to allocate federal agricultural research funds among the states, (c) regional differences in funding patterns, and (d) changes in federal-state relations in funding agricultural research. The paper will then examine policy issues in research funding. These issues will include (a) the potential effect of changes in federal allocation mechanisms on state funding for research, (b) the appropriate role for alternative research funding mechanisms, and (c) broader changes intergovernmental relations that could effect research funding.

Chapter 8: The Trends and Market for Agricultural R&D, Available Scientists, and New Scientists
Wallace E. Huffman

In the United States, agricultural research is comprised of public and private sector activities. Since the early 1950s, the private sector has accounted for more than half of the total research, and the private sector has had a higher rate of growth of real research expenditures over all subperiods in the post-War II period including the 1980s. This is a major factor behind the growing share of doctorates in agricultural sciences being employed in the private sector.

For the period 1980-91, annual data are presented on the supply of new doctorates awarded by field of science for 13 different agricultural fields and for 19 fields in agriculturally related general and pretechnology sciences. In 1980, about three times as many doctorates were awarded in these general and pretechnology sciences as in agricultural sciences. Furthermore, the trend growth in doctorates awarded in agriculturally related pretechnology and general sciences has been much larger than in the agricultural sciences over this period. In particular, during the 1980s, there was a large increase in the doctorates awarded in the fields supporting the new biotechnology, i.e., molecular biology and cell biology.

Chapter 9: Environmental Performance of Agricultural Chemistry: The Role of FIFRA

Robert D. Weaver

By any definition, regulation of economic activity constrains and directs that activity toward chosen goals. As an alternative, tinkering with market incentives can also redirect economic activity by redefining what is profitable. Either approach incurs indirect costs in both the short- and long-run. As has become increasingly apparent in many of the world's economies, these costs may be substantial. As profitability is redefined, the mix of the supply of goods from current technologies will shift, redefining employment of labor and other resources as well as the incomes and returns to those inputs and technologies. Of equal import, as expectations of profitability in the longer term are redefined, the incentives and opportunities for innovation are directly altered and the economy is set off onto a fundamentally different course. Going one step further, in a world of multiple, competing economies, the competitiveness of the regulated or controlled economy is altered.

The regulation of environmental performance of agricultural chemistry is a major force affecting both private and public sector research activity, this regulation has fundamentally reshaped the level of effort and directions of basic as well as applied research, the nature and scale of innovations pursued, and the profitability of research management decisions. This paper presents an economist's view of FIFRA as a regulatory strategy directed toward altering the environmental performance of agricultural chemistry. The paper will reassess the need for and means for redirecting research activity to achieve goals defined through social and political processes. The paper argues that while FIFRA regulation may be effective in altering the environmental performance of agricultural chemistry, ultimately its failure to provide incentives for research that achieves societal goals (e.g. production of environmental impact data) implies that the strategy bridles the potential of U.S. research capacity and the welfare it could generate.

Chapter 10: APHIS and Agricultural Research

Rod Townsend

Federal oversight of genetically modified plants by USDA-APHIS (Agency) under the jurisdiction of the Plant Pest Act (Act) serves as a useful model for examining the impact of regulations on agricultural research in both the public and private sectors. The Act has been criticized on scientific and policy grounds, but nevertheless has provided a functional umbrella for the research activities of those wishing to conduct small scale field trials with transgenic plants. The Agency policy of public notification, coupled with mandated state notification and review, has served to promote public awareness of safe research activities and has proved a major disincentive for states to promulgate their own regulations in this area. To date, 323 permits have been granted under the Act, but only 59 were issued to university or ARS scientists. The logistics and expense of making applications may have inhibited academic researchers, but the system is now routine and costs are minimal. Proposed changes aimed at implementing notification and performance based standards for trials of certain transgenic crops will simplify the process still further and reduce costs to a minimum. In seeking to deregulate further, we must remain sensitive to the issue of public confidence which is fundamental to the commercial success of genetically modified products and the continued support of biotechnology research.

Chapter 11: Risk and Reason: An Assessment of APHIS
Susanne L. Huttner

Biotechnology has been heralded as an important economic development vector for the 21st century. The rate of development of new knowledge and tools has been remarkable -- especially in biomedicine, where clinical scientists have already proceeded with the first trials of human gene therapies. Exciting benchmarks have also been achieved in agricultural sciences. Overall, however, the rate of development has been substantially slower in agriculture than in biomedicine. Perhaps moreso than for other technologies, the rate and directions of seminal biotechnology developments have been strongly affected by governmental actions. Certainly to an extent previously unseen in life science research, federal policies have been determining how some of the very best investigative tools are used in research and training. The difference between the biomedical and agricultural sectors can be traced in part to the federal science and technology policy development process. The worthy goal of the coordinated federal approach to biotechnology policy -- to limit potential hazards while fostering research and economic development -- has proven difficult to attain. There is a critical need to reassess federal policy, especially as it affects basic research. There is a need to achieve greater balance between governmental regulation and incentives, so that public benefit from federally funded research in agricultural biotechnology is not deferred.

Chapter 12: Intellectual Property Laws and Their Impact on Agricultural Research
William H. Duffey

Intellectual property laws and agricultural research are very much intertwined. In the United States there are four categories of intellectual property protection for products of agricultural research, i.e. Utility Patents; Plant Patents; Plant Variety Protection Certificates; and Trade Secrets. This paper examines the role of Utility Patents in agricultural research in America; the protection of products of agricultural biotechnology; and the economic impact of intellectual property protection on U.S. Agriculture.

Chapter 13: Intellectual Property Rights in the Public Interest
Richard K. Perrin

In the past twenty years, intellectual property rights have been extended to crop varieties and other innovations that are embodied in living materials. The nature and extent of these property rights continues to evolve, amid some controversy. These property rights are of intense interest to private innovators but their legitimacy as a public policy derives from their service to the public at large, not just the innovating community. Issues regarding the appropriate strength and breadth of these rights should be examined and resolved on the basis of that public interest, rather than from the more limited perspectives of innovators on the one hand, or potential users of innovations on the other.

This paper describes in an analytical way how the strength and breadth of intellectual property rights can be adjusted to balance the competing objectives of providing incentives for innovating versus insuring diffusion of the innovations. From this perspective, the property rights created by the U.S. Plant Variety Protection Act seem to have been sufficiently strong to provide adequate stimulation of private R&D efforts, yet the royalties charged do not appear to have limited diffusion much.

Chapter 14: Modifications in Intellectual Property Rights Law and Effects on Agricultural Research

W. Lesser

Two aspects of intellectual property rights (IPR) are examined, the effort by developed countries to enhance IPRs in association with GATT, and the 1991 UPOV Convention which introduces dependence into Plant Breeders' Rights (PBR). The focus is on implications for agricultural research. IPRs provide positive if generally weak incentives for private R&D, except for easily copied inventions like plants and pharmaceuticals where they are significant. Enhancing developing country IPR laws is expected to enhance private investments. Due to the localized nature of plant breeding, little of that expenditure will be within the U.S.

Chapter 15: Research Collaboration with the Third World: Has the Free Flow of Genetic material Come to an End?

Wolfgang E. Siebeck

Like national agricultural research, international agricultural research draws heavily on the cooperation of all parts of the global research community. The free exchange of information and of germplasm has served it well, the Green Revolution being but one example. This is rapidly changing: While plant variety protection laws and treaties largely allow scientists to operate by the old rules, patent protection no longer does. This may not all be detrimental. Positive spin-offs should be expected.

A very serious concern is the response of the Third World. When developing countries agreed that their germplasm resources were part of the heritage of mankind, their demands for compensation were ignored. Under the Convention of Biological diversity recently negotiated in Rio de Janeiro (and signed by all attending industrial and developing countries with the sole exception of the United States), their sovereign rights over their germplasm resources have been recognized. This is likely to lend more weight to their claims for compensation, but more importantly, could severely hamper the flow of germplasm and international research cooperation, to the detriment of both developing and industrial countries.

Chapter 16: Changes in Agriculture and Agricultural Institutions
Donald Holt

The mechanism and procedure by which an agricultural research program is funded and managed and the incentives it provides to participating scientists are important in determining the success of the program. The conventional wisdom is that federally sponsored research programs should be funded through competitive grant programs. This raises a number of interesting research management issues.

Chapter 17: Peer Review, Pork, and Priorities in Agricultural Research
Daryl E. Chubin

Formula funding has been the primary Federal mechanism for supporting agricultural research. In the last few years, pressure has been exerted on USDA to adopt a more nationally-competitive, merit-based system. Ironically, this comes when the two R&D agencies for whom project selection has historically been synonymous with poor review, NSF and NIH, seem more willing to experiment with "best practice" and accord greater discretion to program managers.

Chapter 18: The National Research Initiative: Competing at the Margin
Susan Offutt

The groundwork for a significant expansion of competitive agricultural research grants has been laid with the establishment of the National Research Initiative (NRI). Promoted by the National Research Council's Board on Agriculture and embraced by the Bush Administration, the Congressional appropriation for competitive grants has risen from about $40 million to almost $100 million. But the prospects for continued expansion are clouded by constraints on Federal spending and by competition for additional resources from Congressional earmarks of special grants directed to specific projects and states. A comparison of the recipients, size, and subject matter of NRI grants to earmarks reveals great divergence in research priorities and the identify of awardees. The mechanism by which research is funded therefore does matter, and continuation of current patterns may jeopardize the vitality of the agricultural science base.

Chapter 19: Expanding Agricultural Research Agendas with a Shrinking Resource Base: Current Reality for the Public Sector
Katherine Reichelderfer

This paper discusses a strategic approach to improving the performance of the public agricultural research system, which relies on utilizing existing or diminishing levels of resources to expand the clientele base served by the system. This resource-constrained approach recognizes that adjustments towards a downsized agricultural research enterprise will likely be the rule rather than the exception in coming years. Thus, traditional methods of agenda change through expansion of program size are not feasible in the short run.

Chapter 20: Agricultural Research Structures in a Changing World
Brian Wright and David Zilberman

Society's demands on the agricultural research system are evolving from preoccupation with the yield and cost of individual products to concern with safety, quality and variety on the one hand, and environmental implications of production processes on the other. The system's response to the demands will be profoundly affected by the revolutions in biotechnology, ecology and legal protection of agricultural research property rights. The scope of the public role, as exemplified in land grant universities, will be reduced in some areas, expanded in others. New incentives are created by opportunities to sell or license research products under patent protection. The managerial challenge for universities is to use these new incentives to improve overall research performance without compromising teaching, advising and other beneficial scholarly obligations with less direct financial rewards.

Chapter 21: Strategies for Change: Agricultural Research in the U.S. Public Universities

William B. Lacy

Since World War II, U.S. universities increasingly have utilized a rich variety of interdisciplinary centers, institutes, laboratories, research parks, and corporations operating outside the traditional departmental structure to pursue a wide range of new research agendas. As the environment outside the university is shifting and the nature of science and engineering research within the university is changing, university leadership is being increasingly challenged in its endeavor to maintain excellence within the current academic research enterprise. New centers, institutes, and research parks are being formed with primary focus on the major growth areas including information technology and agricultural biotechnology. These new organizational structures, however, need to be larger and better financed than their predecessors. In addition, they are being established and operated by mixed partnerships among U.S. universities, federal and state organizations, private corporations, and private non-profit organizations. Moreover, they are being linked to national, regional, state and local economic development policies with increasing expectations that these new research organizations will provide scientific and technological breakthroughs in key commercial areas. Several examples of new institutional innovations and strategies to promote multidisciplinary or interdisciplinary agricultural research are examined for their strengths and weaknesses.

Chapter 22: Institutional Innovations and Strategies for Public Supported Agricultural Research

Paul F. O'Connell

The Agricultural Research Community needs to revisit it's roots if it hopes to obtain the support required to carry out their scientific studies. That support depends on user relevance, feedback mechanisms, awareness of emerging issues, and most important a true partnership with the final consumer of the research product. Many farmers, commodity groups, agro-industry and public interest groups cannot identify with the current research direction.

Chapter 23: Strategic Issues Facing the U.S. Agricultural Research System
Robert D. Weaver

The chapters of this volume examine from various perspectives the existence and nature of a *general paradox of research* as it may currently affect agricultural research. This paradox is defined by the continued existence of significant calamities that plague society despite substantial achievements claimed by past research and development activities. The chapters add clarity to the view that the U.S. agricultural research system is currently under substantial challenge. Broadly, three types of challenges were identified that face the system: 1) changes in the structure and focus of demands placed on the system, 2) changes in the nature and extent of research opportunity (as defined by scientific opportunity, regulation, and economic incentives including intellectual property rights), and 3) changes in the level and nature of funding of agricultural research.

This chapter summarizes the salient observations of the forgoing chapters and uses them to consider the sources of stress on and nature of challenges faced by the U.S. agricultural research system. The chapter considers these forces within the context of an economic framework that views the level of research effort and support for that effort as being determined through the interaction of societal demands for and supply of research solutions. The chapter concludes that while numerous symptoms of the stress on the agricultural research system may be cited, two fundamental paradoxes continue to lie at the root of declining societal support for agricultural research: *the paradox of success* and *the paradox of relevance*. In brief, the paradox of success is that as the research system has succeeded in achieving increases in quantitative productivity, those successes have weakened support for further quantitative productivity research. The paradox of relevance is interdependent. Despite success in quantitative productivity research, traditional participants in agricultural research has increasingly lost society's franchise to solve research problems concerned with externalities and qualitative productivity of agricultural activities.

About the Agricultural Research Institute

The Agricultural Research Institute (ARI) is an association of industry, academic, government, scientific, and professional groups involved in research related to food, agriculture, natural resources and the environment.

The ARI was organized in 1951 by industrial scientists as a supporting institution for the Agricultural Board of the National Academy of Sciences/National Research Council (NAS/NRC) to identify and promote the kinds of research and policies needed to ensure the best long-term utilization and management of agricultural resources for the national welfare.

In 1973, ARI became an independent, non-profit corporation composed of public and private organizations dedicated to the encouragement and support of agricultural research. It has continued its association with the Board of Agriculture, successor to the Agricultural Board, NAS/NRC.

The ARI provides a forum for discussion of problems common to agricultural research scientists in government, industry, the universities, and other public and private organizations. It serves as a medium for bringing together managers and administrators of agricultural research in the public and private sectors to identify emerging problems and exchange ideas for solving them. It provides a means for initiating actions to implement proposed solutions of these problems.

The purposes of ARI are to facilitate liaison among federal, state, industrial and other groups engaged in agricultural research in its broadest sense; to promote quality, integrity and high standards in the conduct of agricultural research; to define, analyze, and publicize specific problem areas that can be addressed through agricultural research; and to make informed recommendations and disseminate timely information relating to agricultural research in the public interest.

National and international organizations are eligible for membership in ARI, including agricultural research institutions at all levels, federal agencies, state agricultural experiment stations, academic institutions, philanthropic foundations, scientific societies, corporations, organizations and individuals having an interest in research, development, production, processing, transportation, distribution, and marketing of agricultural products and agricultural services, or in manufacture of supplies and equipment used by the agriculture community.

The ARI holds conferences, workshops, and an annual meeting to transact business and present a technical program, the proceedings of which are published. New members are most welcome.

The Agricultural Research Institute • 9650 Rockville Pike • Bethesda, MD 20814
Phone: 301-530-7122 • FAX: 301-530-7007

Sustaining Members - FY 1992

Demonstrating strong leadership in support of the Agricultural Research Institute and in the promotion of agricultural research for the betterment of mankind, the following organizations have joined the Institute as Sustaining Members:

American Cyanamid Company, Agricultural Research Division
American Society of Agronomy
Anderson Foundation
Campbell Soup Company
CIBA-GEIBY Corporation, Agricultural Division
E.I. duPont de Nemours & Co., Inc.
FMC Corporation, Agricultural Chemicals Group
The University of Georgia Agricultural Experiment Station
ICI Americas, Inc.
Langston University, Agricultural Research and Extension Programs
Merck Sharpe & Dohme Research Laboratories
Michigan State University Agricultural Experiment Station
Mobay Corporation
Monsanto Agricultural Company
National Agricultural Chemicals Association
Ohio Agricultural Research and Development Center
The Pennsylvania State University Agricultural Experiment Station
Pioneer Hi-Bred International, Inc.
Pitman-Moore, Inc.
Purdue University Agricultural Experiment Station
Ralston Purina Company
Rhone-Poulenc Ag Company
Tennessee State University
United States Environmental Protection Agency
Virginia State University

RH (eds) Deformation of the continental crust: the legacy of Mike Coward, vol 272. Geological Society, London, Special Publications, pp 155–185

Upadhyay R, Frisch W, Siebel W (2008) Tectonic implications of new U-Pb zircon ages of the Ladakh batholith, Indus suture zone, northwest Himalaya, India. Terra Nova 20:309–317. https://doi.org/10.1111/j.1365-3121.2008.00822

Weinberg RF, Dunlap WJ (2000) Growth and deformation of the Ladakh Batholith, Northwest Himalayas: implications for timing of continental collision and origin of calc-alkaline batholiths. J Geol 108:303–320